THE *ART* OF
WOODGRAINING

THE *ART* OF
WOODGRAINING

STUART SPENCER

Macdonald Illustrated

A Macdonald Illustrated BOOK

Text copyright © Stuart Spencer 1989
Illustrations copyright © Macdonald Orbis
1989

First Published in Great Britain in 1989 by
Macdonald & Co (Publishers) Ltd
London & Sydney

a member of Maxwell Macmillan Pergamon
Publishing Corporation
Reprinted 1990

British Library Cataloguing in
Publication Data
Spencer, Stuart
The art of wood-graining.
1. Graining. Manuals
I. Title
745.51

ISBN 0-356-17536-7

Filmset by Flairplan

Printed and bound in Italy by Graphicom

Senior Editor: Judith More
Text Editors: Jonathan Hilton, John
Wainwright, Philip Wilkinson
Senior Art Editor: Clive Hayball
Designer: David Rowley
Photography: Jerry Tubby, Martin Cameron,
Susanna Price
Picture Research: Kathy Lockley
Production: Jackie Kernaghan
Illustrator: Paul Richardson
Macdonald & Co (Publishers) Ltd

Orbit House,
1 New Fetter Lane,
London EC4A 1AR.

CONTENTS

INTRODUCTION

The diversity of pattern and colour found in many timbers is of such stunning variety and complexity that, throughout the ages, both artists and craftsmen have been inspired in the pursuit of the painted illusion. Wood is mankind's oldest and most cherished material. As well as providing fuel and shelter, it has been used in the manufacture of a range of goods so vast that for many thousands of years it has fundamentally defined the material world in which we live.

Although the basic process of development and growth is common to all trees, natural forces and local conditions produce an extraordinary diversification of colour and pattern within the various species. When converted to timber, whether for utilitarian or decorative purposes, the internal structure of the tree is revealed in the form of a series of markings, or figuring patterns. These explicit signatures are more than superficial cyphers, they are a statement of growth and development and an indication of the strength and performance of the individual timbers.

Our respect and admiration for wood is assured. Endearing references to wood in displaying such disparate qualities as warmth, honesty, charm, elegance, and sophistication indicate something of the potential that may be implied in the grained surface. Our understanding of these features is so instinctive that the crudest attempts at imitation trigger not only instant recognition but also a deep-seated, intuitive response. In this respect, the strength of any painted illusion will be judged not only on the quality of the technique but also on the sympathetic manner in which it is incorporated into the domestic environment ∎

The doors have been grained with a limed oak technique, and the characteristic markings of the fractured 'flake figures' are clearly evident in the finish. The choice of oak confirms the formality of the setting, with the paired arrangement providing both balance and continuity, while a gilded dado rail helps to define the pleasing proportions, and emphasises the quality of the decor.

This book starts with 'Understanding wood' — a brief description, with accompanying illustrations, of the growth and 'harvesting' of genuine wood, to give a background to the terms used later. Next, chapters Two and Three show how woodgraining can be incorporated sympathetically into a wide range of home design schemes, from the traditional English look to post-modern style. Then three practical chapters — creating panels and panelling systems, preparing surfaces and choosing and using paint, glazes and varnishes — follow. Finally, the techniques chapter reveals the secrets of woodgraining methods, with detailed descriptions of the basic techniques and illustrated step-by-step instructions to enable you to reproduce 23 different figures from 8 different woods.

Technique is a word that crops up many times in this book. You can regard technique as an acceptable short cut or a time-saver in imitating specific qualities that may be found in the natural material. Whether this involves the ingenious use of some specific property of paint or the adaptation of some 'found' tool or brush, it is a way of automating a process, which, over large areas, could become both tedious and time-consuming. Techniques are in fact 'tricks of the trade' and, understandably, professionals have for centuries

safeguarded their livelihood by shrouding them in tradition and mystique.

In the techniques chapter you will see the simplicity of the various processes involved in the simulation of wood, but it must be stressed that it is only through practice that you will achieve professional results. In order to secure the illusion, mastery of technique and confidence with colour must be combined with a clear understanding of the natural material and the manner in which it might be employed.

Although flat paint has always provided the most effective way of decorating homes, the opaque nature of its composition severely restricts its potential as a decorative finish. The use of transparent colour in the form of glazes and washes is fundamental to the illusory arts. In this book you will see how they can be used to extend the language of colour in the home.

Historically, woodgraining has been used by all levels of society, but it is in the prestigious municipal buildings and homes of the wealthy and privileged that we discover the most arresting examples. These often take pride of place among surfaces and artefacts that are certainly authentic and valuable. Why should the painted illusion be incorporated in situations where quality is paramount and cost hardly a limiting factor? The answer is to be found in the phrase 'creative interpretation'. This quality is the essential ingredient that elevates what could be little more than mere mechanical imitation into a painted expression of great subtlety, charm, humour and wit.

A final word of encouragement for those new to decorative painting: although first attempts may often appear awkward, naive and sometimes a little crude, they nevertheless project an immediacy and vitality that is often absent from the more contrived presentations of their slick, professional counterparts. While there is no substitute for experience, enthusiasm, a fresh eye and a sense of humour are all vital ingredients in the pursuit of the painted illusion.

Right: *Peter Farlow has used dark-grained mahogany areas to identify elegant architectural detail. Although sharp lines and strong contrasts proliferate, the colours are warm and there is sufficient distraction in the arched window and angled corner unit to offset any feeling of regimentation that might occur in the confines of this small space. The enclosed panels feature a bird's eye maple simulation, and the mild yellow tones and soft patterns of the wood are pleasingly reflected in the delicately mottled finish of the washed walls.*
Left: *For centuries, woodgraining has been used as a substitute for rare and exotic veneers, even in such prestigious locations as Brighton Pavilion, the English Prince Regent's seaside palace.*

CHAPTER ONE

UNDERSTANDING WOOD

In order to simulate the various decorative characteristics and markings of timber it is first necessary to have a basic understanding of the structure of the tree itself, the way it grows and develops and the manner in which it is ultimately converted into timber.

A fundamental grasp of a few elementary facts and definitions at the outset will prove invaluable when you wish to replicate a particular wood accurately without the assistance of a sample or, indeed, when you wish to produce more abstract interpretations of the figuring characteristics.

Despite the fact that there are many thousands of species and sub-species of trees, they all grow and develop in a similar manner. A root system absorbs minerals and water from the ground and provides a structural base to support the trunk. The trunk carries the branches and provides a transport and food-storage system for the various fluids that pass through the tree. The branches carry the foliage to the sunlight and the individual leaves provide food for the tree by a process known as photosynthesis. Chlorophyll, a substance present in all green leaves, permits ultraviolet light from the sun to combine with carbon dioxide from the atmosphere to produce the sugars, or food, necessary to sustain the growth of the tree.

The many species that exist on the planet result from the varying natural constraints imposed by such factors as location, climate and soil condition ■

The generous proportions of the manor house comfortably accomodate the dark presence of this traditional oak staircase. A harmonious balance is achieved within the interior by contrasting the sombre, understated patterns and colours of the antiqued oak finish with the lighter panelled soffit, striped walls and gilt frame.

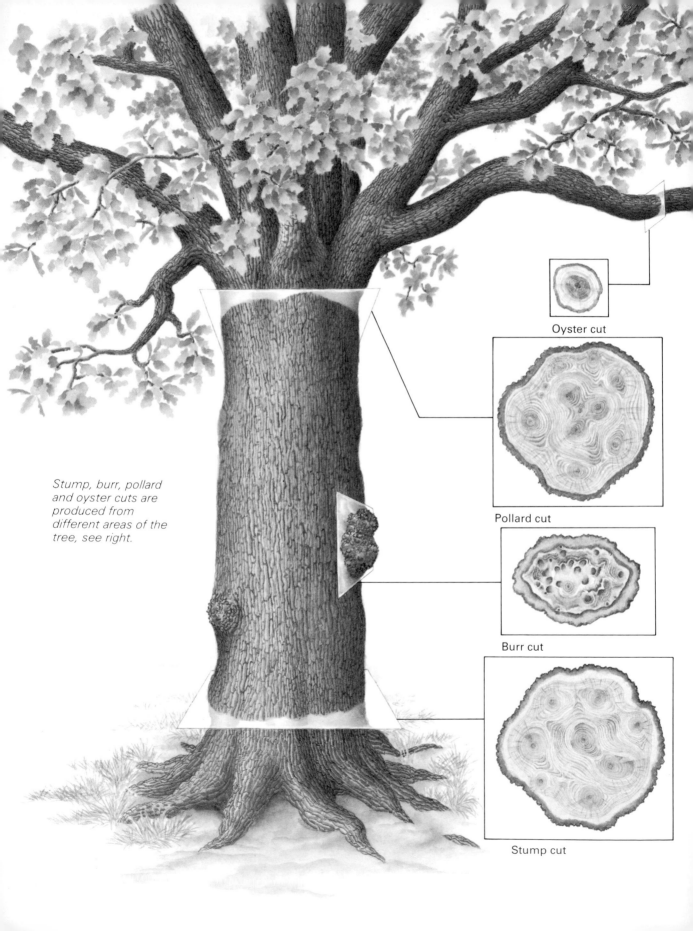

Stump, burr, pollard and oyster cuts are produced from different areas of the tree, see right.

Oyster cut

Pollard cut

Burr cut

Stump cut

The structure of the tree

A tree is a living organism composed of millions of cells or fibres. These are organized mostly in a vertical direction to provide strength and elasticity and to conduct essential fluids throughout the body of the tree.

Softwood cells tend to be geometrically organized. In hardwoods, cell organization can vary considerably but usually the body of the tree is made up of long, fibrous cells with thick walls that produce a visually more textured appearance. In addition, vertical cells or vessels, known as pores, can also be seen. These provide an easy way of differentiating hard from softwoods.

The directional organization of the combined fibres in the trunk is referred to as the grain and it is clearly visible in the majority of timbers. While grain is organized in a predominantly vertical direction, variations in the form of wavy or interlocked grain produce some of the most decorative figuring.

Annular rings

If you take a horizontal cut across the trunk of a tree you will see the roughly concentric bands of the annular growth rings. Each ring represents the new wood that is formed on the outside of a tree in one growing season. The growth of the ring is not uniform;

the less dense and lighter-coloured wood is produced first, while the darker, more dense wood occurs in the latter part of the growing season. Branch and root systems originate from the pith or centre of the tree. They grow and develop in a similar manner to the trunk and show the same annular ring formations. Knot formations indicate the positions of branches, and close inspection will reveal the tight concentric banded patterning of the annular growth rings.

Sapwood and heartwood

The trunks of many mature trees show a progressive darkening towards the centre. This is known as the heartwood and its function is to provide mechanical support. Around the outside of the heartwood is found the lighter-coloured sapwood, which acts as the transportation system for food and water. In some cases there is a distinct delineation in the colour and figuring between the two, and the contrast can be highly decorative.

Rays and knots

Rays are horizontal elements primarily responsible for the transfer of food across the girth of the tree. They are more pronounced in hardwoods and are clearly seen in the characteristic flecks in beech or the silver grain in oak.

Above, the softwood Scotch Pine and below, the hardwood Horse Chestnut, are shown with their respective cone and conker seeds.

Converting logs into timber

The manner in which a log is cut influences its physical performance and, more important for our purposes, its appearance. Timber is cut in one of three ways.

Through and through sawing is also sometimes referred to as tangential or plain sawn. In this process the log is sawn by parallel cuts along the length of the grain. The growth rings in plain sawn timber meet the face of the board at an angle less than 45°.

Quarter sawn is also known as radial cut timber and it produces boards where the grain meets the face of the board at an angle not less than 45°. The process is more expensive, since the log has to be turned to achieve the various cuts. It does, however, have the advantage of producing boards that are more stable when drying out and, again more important for our purposes, it reveals the highly decorative ribbon-like rays that radiate from the heartwood. These rays appear as the prominent silver slivers commonly found in sawn timbers such as oak or beech.

Finally, there is the rotary sawn method; where the trunk of the tree is roughly cylindrical it is economical to rotate the tree and literally peel thin sections of timber veneer from the log in a continuous operation. Veneers produced in this way are generally used in the manufacture of decorative laminates for the furniture industry or as durable plywood composites for the building trade. The boards produced have a continuous patterning with a rough repeat that is a function of the circumference of the tree from which they have been peeled. To the untrained eye the impression given is that the timber has been cut from a tree of enormous girth. Examples are plentiful on building sites. While rotary sawn veneers are essentially utilitarian in nature, their explicit figuring may provide you with some inspiration for the composition of larger-scale painted woodgrain illusions.

The pattern of growth
Figuring patterns in wood are derived from the growth habits of the tree. Each year, circular bands of new wood grow on the outer edge of the trunk, inside the bark. One year's growth is represented by two bands – a light one produced in the early part of the season, and a dark one that develops later in the season.

A section through the trunk reveals the different patterns that result from this growth phenomenon: **1** *The compacted flame figure of the central heartwood.* **2** *The tight pinstripe parallels of the surrounding sapwood.* **3** *The transverse rays that traverse the girth of the tree.*

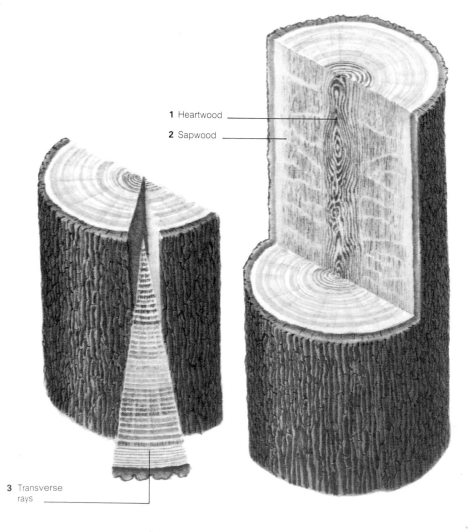

1 Heartwood

2 Sapwood

3 Transverse rays

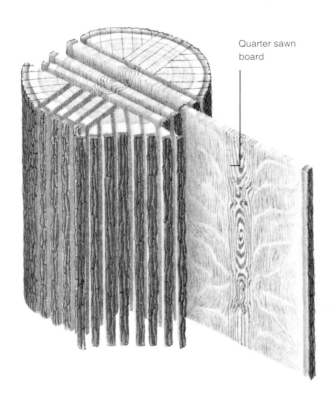

Quarter sawn board

Cutting the log

The way that a log is sawn up determines both the appearance and functional properties of the cut timber or lumber.

For example, the diagram on the right shows the different ways of extracting a tangential *or* quarter sawn *board from a log. Boards produced in this manner tend to be more sturdy and reliable than other cuts, as well as being more decorative. Indeed, it is as the saw cuts through the tangentially radiating 'rays', that the* flake *or* silver grain *figure, characteristic of so many hardwoods, is revealed.*

The second diagram, below left, reveals how plain *or* through and through *sawing methods are used to extract boards from a log. Although small boards taken from the outside sections have a tendency to bow or warp during seasoning, they nevertheless display attractive flame figure markings.*

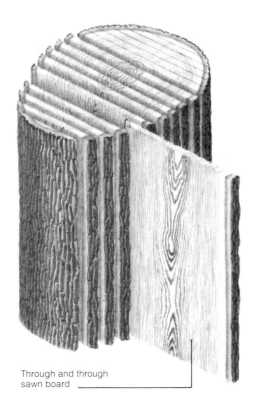

Through and through sawn board

Rotary sawn wood (above)

Although rotary sawn timber is generally employed for utilitarian purposes, the flowing patterns produced by this cut are both graphic and distinctive. And as such they can provide a useful and almost infinite source of inspiration for the production of freehand figuring and fantasy graining.

The figuring of wood

Figuring is a general term applied to the highly prized decorative markings that are evident on the surface of sawn timber. As a novice wood grainer, it is in this section that you may begin to appreciate the endless variety of patterning produced by the natural processes of growth. Although woods belonging to the same species exhibit comparable figuring characteristics, no two cuts will ever be the same. In this respect, when you are trying to produce a particular wood grain effect you must use your experience and imagination to express a typical example.

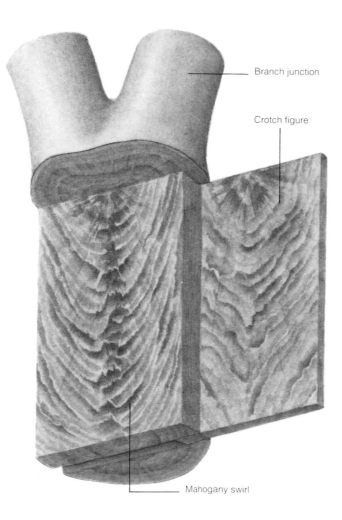

Branch junction

Crotch figure

Mahogany swirl

The distinctive mahogany crotch figure becomes evident when the tree is sectioned through the central section of a branching junction. As the sections retreat from the centre, the figuring becomes less symmetrical, producing the abstract mahogany swirl.

Growth ring figures

Growth ring figures are those that perhaps most people think of as wood grain; they include flame or crown figure and silver grain figure.

Flame or crown figure If the growth and structure of a tree were perfectly regular, when it was made into timber you would see along its length a regular, striped pattern. Since this rarely happens in nature, irregular growth produces a series of undulating bands, which, when intercepted in the cutting process, give distorted, arcing ellipses and parabolas. These are sometimes referred to as crown or flame figures for obvious reasons and are seen best in plain sawn timbers such as oak, elm, chestnut and ash, and to a lesser degree in walnut, mahogany, birch, beech and maple.

Silver grain figure When rays are intercepted in radial sawn timber, shiny, silver flecks appear. These may be randomly organized or they may form distinctive patterns. Australian oak exhibits a fine mesh structure, while European oak displays the 'flake figure'.

Pigment figure

Certain woods produce areas of contrasting colour that bear no reference to the natural growth-ring structure of the tree. This figuring is most evident in Circassian walnut and figured red gum, and it produces a surface of great subtlety and depth.

Curly figure and crotch figure

These figures are produced at the base of the intersection of a large branch with the trunk. A plain sawn section reveals the curly figure and the radial sawn section the crotch figure (see left).

Interlocked grain figures

These features occur in some tropical woods. The grain of annular growth rings twists or spirals in alternate directions up the length of the tree. When it is radially or tangentially sawn, a prominent striped pattern is produced by the reflection of light from the interlocked grain. Mahogany and satinwood express this feature to its fullest advantage. Where the stripe becomes wider it is known as ribbon figuring.

Where forces disrupt the natural growth pattern, the grain is thrown into vertical and horizontal ripples, and when a saw passes through the crests of successive undulations, the normal stripe figure is distorted into a bizarre series of mottled patterns. There are more than twenty of these in mahogany, for example. Typical patterns include:

Fiddle back This describes a series of fractured horizontal lines roughly arranged into undulating parallel stripes. The figure is so named because it is tradition-

In these diagrammatic representations of figuring details, examples of the wide variations that occur within the natural structure of the wood grain pattern are shown. Although figuring patterns are specific to the various species, external influences, such as climatic conditions, horticultural processes, or even parasitic invasion, can produce variations that make the wood a highly-prized decorative surface.

Satinwood – bee's wing figure

Walnut – pigment figure

Mahogany – plum pudding figure

ally displayed on wood used in the manufacture of the bowed surfaces of the backs of musical instruments.

Bee's wing This is a more broken version of fiddle back and its appearance can be likened to the veined formations on an insect's wing.

Block mottle This effect is produced by an irregular zigzag distortion of the stripe figure.

Rope figure In this figure a staggered arrangement of dapples in the stripe appear to twist as they ascend.

Raindrop figure In this example, irregular diagonal mottles take on the appearance of slanting raindrops.

Blister figure This rare marking appears as crushed velvet with broken thread-like formations linking irregular mottled areas of light and shade.

Although the interlocked figures listed here have been identified as separate patterns, in reality sawn sections of timber may exhibit surfaces composed of two or three examples, each merging gradually into the next.

Horizontal figures

These are horizontal sections taken near the base of the tree and at the top where many branches can be seen exiting the trunk simultaneously. Both are taken in the form of thin veneers, the former being referred to as a stump cut, the latter as a pollard cut.

Pollarding is a horticultural process of selective pruning, in which branches are induced to 'bush'. The cuts show evidence of root and branch formations respectively and they appear as small eyelets or pronounced knots surrounded by swirling annular growth rings and heavily pigmented figuring.

Burr figure

Burrs are parasitic growths that sometimes occur on the main trunks or branches of a tree. The formal growth pattern of the tree becomes disturbed and disorganized and the resulting formations provide highly decorative veneers.

Oyster figure

This is the name given to decorative cuts made through the small branches of selected trees, for example laburnum. They reveal the concentric, annular ring formations of the end grain structure.

Bird's eye figure

Sometimes the entire body of the maple is invaded by parasites. Where the cut intersects these imperfections, the eyes appear as small ovals with shadowy interiors. Their relationship to the predominant figuring produces a finish that resembles a diagram of a complex road system.

'Plum pudding' figure

This figure is seen in mahogany and it is produced by chemical impurities that are assimilated within the body of the tree. Although the effect is similar to a bird's eye figure, the spots are elongated or lenticular in the vertical plane.

The use of wood

Although a description of the physical performance of timber falls outside the scope of this book, for the sake of authenticity in your painted illusion it is a good idea to bear in mind certain physical limitations of wood.

Wood shows great strength in compression and tension along its grain. It has reduced tensile strength across the grain and even lower resistance in sheer parallel to the grain. Thus you can see that in the use of wood, the figuring reflects physical performance. This information is valuable when, say, woodgraining a beam where, for example, certain incorrect figuring would read as a visual nonsense.

The use of wood is so extensive that inspiration for the grained signature is all around us. Always be on the look out for interesting formations or colours. A camera or sketch book is handy to record figuring patterns or grain formations that catch your eye. Once you have been seized by the potential of the illusion, the larch lap garden fence, the old pine dining table, the rough hoardings surrounding building sites are all fair game in providing the inspirational material for the next illusion.

Look closely at the three-dimensional wood surface. Carvings, mouldings and structural members contain some of the most exciting and decorative figures. Prestigious buildings have traditionally used timber as a decorative and structural material, with museums, churches and town halls exhibiting quality woods and craftsmanship. Stately homes will show you the more intimate domestic use of wood, and period furniture will provide you with endless examples of the decorative use of timber in parquetry design and wood combinations.

Some of the white softwoods have very absorbent tissue and can be readily stained to resemble more prestigious timbers. Also, in contemporary design, the use of woods that have been stained unnatural colours is becoming increasingly popular.

The natural surface of wood may be protected with polish or varnish. Unprotected woods tend to lighten with age, although successive applications of wax or polish will darken and mellow the surface. While woodgrainers may be called on to simulate age or deterioration in the form of 'antiqued surfaces', on the whole a simulated wood finish will resemble a natural finish that has been lightly polished.

Mahogany grained panelling has been used extensively within this otherwise pale-coloured interior to accentuate the formal geometry of the room, highlight the window recesses, and develop an air of refinement and quality.

CHAPTER TWO

WOODGRAINING AND DESIGN

This section shows how the various woodgraining signatures can inspire the painted illusion and where woodgraining effects can be best used in the home. This information is broken down into three broad areas: visual properties, character, and interpretation — literal or metaphorical?

The visual strength of the natural wood surface is a combination of its overall colour or tonal value, its reflective qualities and the contrast and prominence of the figuring pattern. The ability to orchestrate these features allows the grainer to control the visual presence of a surface: should it visually advance or retreat, for example, becoming either an eye-catching feature or an anonymous backdrop?

If we are to incorporate the woodgrain signature successfully into the home we must understand something of the character of timber. The material is capable of wearing many hats and the painted illusion can be used to good effect in creating or complementing different atmospheres found within the home.

For some, the art of successful woodgraining relies on a detailed knowledge of specific samples and a thorough mastery of the paint mediums and tools. The results in many cases are indistinguishable from the real thing and any deviation from this standard is regarded as sloppy technique. Others see the first approach of photorealism as too rigid and sterile. A free spirit, imagination and a sense of humour are vital ingredients in the armoury of those who wish to create fantasies ∎

The presence of this simple, combined bookcase and cupboard unit has been considerably strengthened by picking out its various sections in contrasting graining techniques. Bird's eye maple panels have been featured within the contrasting borders of a simulated oak burr.

Colour and lustre

While the colour of the natural wood surface embraces nearly the entire spectrum, from whites and pale yellows, through browns and reds to deep purples and inky blacks, on the whole its finish remains subdued and understated. In this respect the natural colours of most woods are easily assimilated into the home.

Tonal variations in colour are common to most woods and produce pleasing, visually textured surfaces, especially in the lighter varieties. Although they often occur around knots or faults in the grain structure, tonal variations may develop independently of the predominant figuring pattern. In extreme cases, discolorations can be pronounced, and examples of these are highly prized for their decorative potential. For example, in the case of yew, reddy brown, undulating, ribbon-like formations, closely resembling the veining features found in marble, are often seen striking out over the creamy white ground between carelessly scattered knots. While burr, root and pollard cuts provide tonal variations that contribute significantly to the form of the figuring in producing the mysterious swirling patterns typical of these sections.

Although some of the unicoloured woods, such as ebony or tulip wood, are among the most sought-after cabinetmaking timbers, generally those that show little or no pattern are set aside for more mundane purposes or they are used in the form of marquetry or parquetry veneers to contrast with the more decorative figuring of other woods.

It is worth bearing in mind at this stage the breadth of taste in the contemporary display of wood surfaces. At one extreme there are those who, recognizing the decorative qualities of wood, proceed to strip all painted timber to reveal its natural features. At the other extreme, there are those who prefer white wood that has been dyed to produce totally unnatural hues, such as bright reds, blues and greens. The figuring and grain structure are then revealed in a slightly contrasting tone to the predominant colour. While the results are very much an acquired taste, it is worth considering exploiting this phenomenon, especially in view of its popularity within the contemporary interior.

Lustre or sheen is the reflective quality exhibited by many of the highly prized tropical hardwoods. It is a natural feature produced by interlocking grain formations and it is independent of any varnishes or polishes that might subsequently be applied. With this type of timber, as you move your viewpoint, the surface appears to ripple or shimmer. This can occur in isolated spots or you may be able to see it as a regular repeat. In the latter case, you can construct dominant patterns by creating accents in the mottled figuring (see *Static repeats*, page 24).

Reflective surfaces tend to be formal and sophisticated. Although they visually advance, the effect is neutralized by the qualities of movement and depth implied in their surfaces. Your success as a grainer in replicating these features depends on your ability to mix tonal variations of the same colour and your skill in handling a mottling brush. Surfaces can be constructed as backdrops or as features and you can adjust the scale of the mottle repeat to suit small objects or large areas.

Left: *Adam Arbeid combined marbled and grained surfaces in this pleasing, painted fantasy. The individual sections of the door (rails, stiles, panels and frames) have all been clearly identified by slight variations in technique, and subtle changes of shading.*

Right: *Mahogany graining by Michael Synder has given this corner storage unit an elegant presence. The plume of the inverted crotch figuring has been used to identify the various door and drawer facings. Plainer, striped figuring elsewhere provides an effective contrast to the decorative sections. The pale-coloured backboard lightens the piece.*

Figuring

The visual strength of the surface is a combination of the directional strength of the figuring pattern and the degree of contrast it exhibits between itself and the overall tonal value or background colour. By controlling these features you are able to control the movement of the eye over the grained surface. In this respect the illusion offers great potential, especially when you are treating large areas within the home.

Insistent direction

In plain sawn timber the annular growth rings appear as a series of vertical parallels. Depending on the type of timber, these may vary between fine, regular pin stripes and broad, undulating bands. Where strong contrasts occur the effect is strident, especially over large areas. More decorative cuts are produced when the saw meets the stripe at a tangent and the characteristic arcing ellipses and parabolas of the crown figure are formed. These patterns are visually compelling; the eye is directed over the surface. In replicating these variations, you can exercise a lot of control over the visual strength of surfaces. (On vertical surfaces such as walls or doors, the figuring is constructed in a rectilinear fashion as a vertical or horizontal accent.) In order to retain an authentic feel for the material where large areas are being grained, the simulation is often displayed in the form of a panel system. A repetitive design accentuates and exaggerates the vertical or horizontal accent and you can use the illusion to good effect in emphasizing or disguising good and bad proportions within a room (see *Panels and panelling systems,* pages 50–7).

Although mottle features tend to be understated, you can produce strong rectilinear and diagonal accents in figuring large areas. Conversely, where it is necessary to tone down an overinsistent figuring pattern, introduce contrasting elements or features, such as mottles, prominent knots or silver grain, to reduce the impact. These elements draw or distract the eye from the predominant accent of the background figure.

Static pattern repeats

Examples of this type of figuring include the blister, fiddle back, block mottle and plum pudding figures (see pages 16–17). Although these finishes exhibit tight, repetitive themes, they are not insistent and the eye is invited to browse over an evenly textured surface. Static patterns may be relieved by adjusting the strength and direction of mottle features. In reality, large areas of consistent pattern repeats are rare, since the natural material gradually transmutes to some less clearly stated figure.

Focus patterns

As the name implies, the eye is focused on a central feature. This may be small details within a surface or large featured panels.

Bull's eyes The most definitive focus pattern is the 'bull's eye' of the oyster figure. Oysters are small, between one and four inches in diameter, and are generally combined in groups to form simple geometric designs on elegant furniture. Individual oysters may display perfect, concentric circles or distorted ellipses.

Burr, root or swirl figures These display intricate and detailed random patterns, drawing the eye into the depths of swirling, mysterious vortices and whirlpools. The majority of trees exhibit these complex grain features, but they are most pronounced in walnut, yew, oak and ash. Although the pattern density is very small, upgrading the scale can once again produce eyecatching fantasies.

Crotch and swirl figures In rare cases grain patterns and growth ring figures produce features of such striking size and symmetry that they are displayed as composed designs. The most prized is the crotch figure – a radial cut taken at the base of the intersection of a major tree limb and the trunk (see page 16). Compared with the way it is naturally produced, the pattern is displayed on vertical surfaces in an inverted form. The swirl figure produced at major branch sections can also be included in this group.

Combination patterns

Traditionally the colours and figuring characteristics of the various woods are combined in a variety of decorative ways to produce balanced geometric designs.

When cutting wood into veneer form, the slow pattern progression through successive slices allows the apparent mirror imaging of the grain patterning to produce book-end or double book-end designs. The possibilities for marquetry and parquetry on furniture are endless and the technique may be used to great advantage in focusing or moving the eye.

Although similar book-end patterning techniques can be used on a larger scale in the production of panelling systems on vertical and horizontal surfaces, the traditional formula relies on the use of rails and stiles to outline the various panels. In woodgraining, the process of rendering these features is known as quartering. The grain on rail and stile is generally expressed as a vertical or horizontal accent.

Simple combing and dragging techniques were used to grain these units. Soft tonal variations and delicate patterning prevent them from dominating the room.

A natural or a polished look?

Pine is a functional and unpretentious timber, unsophisticated, sometimes crude but always an honest surface that displays a rustic simplicity, warmth and charm. These characteristics are obviously more desirable in a kitchen or child's bedroom than the refined atmosphere of an elegant drawing room. For example, the traditional pine table is a popular piece of furniture found in many kitchens. Twisted boards, knife cuts, stains and burns are part of its appeal and bear witness to its utilitarian role. The structure tends to be simple, with little obvious attempt to organize consciously the patterned surfaces or elements into any prescribed decorative pattern. Unprotected by wax or varnish, the wood has a washed-out appearance and the light gray to fawn colouring is often toned with vague red, green or purple discolorations. The mid-brown figuring is strong but never strident, with attractive crown and flame formations linking numerous and randomly positioned knots.

Over the centuries oak has established itself as the king of woods. It is stable, durable, easily worked and extremely decorative. Although the wood is predominantly a rich brown colour, there are many varieties and the tone may vary from a light tan through to almost black. Quarter sawn oak planks display the beautiful flake figure, while plain sawn boards carry

Above: *The warm, mellow tones of pine are clearly evident in Peter Farlow's simulation, shown in this detail, and provide an understated backdrop to a framed picture.*

lively and attractive flame and crown markings.

Although simple, a traditional oak table will generally reflect the hand of the craftsman in design and construction. It is a robust, unpretentious and hearty wood that is valued for its warmth and integrity. The surface is waxed rather than varnished in order to reveal its mellow characteristics, and its presence is welcomed in any situation from a cosy parlour to a grand room.

By contrast, the formal hardwoods and veneers used for classic furniture, such as mahogany, display a quality and elegance that is evident in the form, materials and construction used for pieces. On a mahogany table, for example, the structure is clearly defined as base and table top and these elements are generally expressed in matching but differently figured examples of the same wood. The woods or surface veneers are positioned so that the selected figuring patterns or parquetry designs flatter, complement or reflect the natural form of the piece. Varnish and polish nurture, protect and enrich the natural colouring and the table reflects a cherished, cared-for appearance. Refined, subtle, dignified, formal, and sometimes even ostentatious are all adjectives that can be used to describe such a piece. Certainly these characteristics would indicate its correct location is in an elegant drawing room rather than a simple kitchen.

Left: *The coarse texture of Peter Farlow's limed oak simulation is complemented by the polished chair. The detail (above) shows the ragged serrations of heartwood and the overlapping format of the flake figure.*

Authenticity or fantasy?

There exists between the two extremes of photorealism and abstraction sufficient scope to accommodate all levels of taste and performance, although practice and familiarity with the techniques will determine where your aspirations and preferences lie. While you may balk at the dexterity, draughtsmanship and precision of the literal interpretation, and be seduced by the freedom of abstraction, the personal statement can often prove an elusive goal. The creative process is demanding and the search for originality can be full of frustration and disappointment.

Often, you will find that the examples which linger in the memory are those that steer dangerously close to recognition itself and yet still manage to retain the vital essence of timber. An illusion that is only just an illusion – that is almost dismissed as too naive and simplistic, yet possessing sufficient subtlety to draw the response. It is the inquisitive yet disbelieving eye and the smile of recognition that are the ultimate salutes to the skill of the artist.

Whichever approach you adopt, remember that decorative painting is a means to an end rather than an end in itself. To view your efforts in isolation is to tell only half the story. The responsibility of the decorative painter extends to the whole visual environment and the sensitivity with which the illusion is incorporated in the home is every bit as important as the detailed nature of the simulation itself.

Historical perspective

In the latter part of the nineteenth century the decorative arts in Europe had undergone a revival. The criteria for success were based on the ability of the decorative painter to replicate with photographic precision the many nuances of the various species of timber that had gained favour among the cabinetmakers of the day. The results speak for themselves: so authentic were the finishes that a magnifying glass was required to differentiate the simulation from the genuine example. It was at this time that a recognized school of decorative painting was established in England, which featured John Kershaw and other leading lights among its followers. In recognition of the skill and commitment of these great technicians, the Victoria and Albert Museum and the Bolton Museum now display examples of their work.

Contemporary settings

Although the origins of decorative painting are rooted deep in tradition, there are no formal rules to dictate the manner in which the discipline should be interpreted or in fact those areas where it should most fittingly be applied. If the techniques are to have any relevance in a contemporary setting they must reflect the mood and style of the moment and, where necessary, complement the fashions of the day.

In the past, 'good taste' dictated that graining should be restricted to those surfaces or objects that would have naturally employed or displayed the authentic material. Values change and recently this philosophy has been replaced by a more enlightened approach, in which all manner of artefacts may be considered as suitable subject matter. Fantasies exploiting the visual joke have become very much a popular format for the expression of the illusory arts.

Left: *Painted here by John Kershaw, perhaps the most popular of all decorative figures is the mahogany crotch. The beautiful symmetry revealed in the spray or plume has, for centuries, inspired painters.*

Right: *In this minimalist fantasy grained surfaces are intentionally crude, and the figured patterns little more than exaggerated cyphers. They are complemented by painted tapestries of a similar tonal value.*

CHAPTER THREE

WOODGRAINING IN THE HOME

It is only by incorporating your efforts within the home environment that you can develop the full potential of the woodgrain illusion. Few of us are fortunate enough to begin with a blank canvas, and a measure of your skill is the success with which you can combine the illusion with existing fixtures and fittings. However, the traditional application of wood in the home ensures that there is plenty of scope for you to integrate realistic simulations.

If you have an adventurous approach, you must bear in mind that graining is a means to an end rather than an end in itself. The interpretation is completely subjective, and contemporary themes may demand far more abstracted solutions in the way that colours, colour contrasts and figuring are represented.

A brief checklist will help you to organize your approach. Initially, you must identify the surface or object to be grained. Then you must decide on the character or style you want to develop. Character will indicate the various woods that may be used and the decision to prepare surfaces as backdrops or features will suggest the most effective colours, contrasts and figuring patterns. Finally, let the scale and proportion of the room determine the display format and dimensions of grained areas in terms of panel systems, parquetry designs or featured areas ∎

Hannerley Dehn has used the typically disjointed figuring of rosewood to counterbalance the rigid geometry of the piece and she has grained a fine contrasting border, in a plainer stripe, to redefine the outer edges. When inspected at close quarters, the bizarre patterning of rosewood is mesmeric and demands attention.

Halls, stairs and landings

Halls, stairs and landings link the various rooms and functions within the home. But because people move through rather than linger in these areas, they tend to be dismissed as being incidental to the overall design theme. This is a great oversight, and if properly handled these 'flow areas' can be developed as one of the most dramatic interior decoration themes.

Remember, first impressions count. While the family may take these areas very much for granted, for visitors the quality and style of the decor they initially encounter can affect the way they regard the rest of your home. Also these areas link not only the different functions but also the different styles of the various rooms. Continuity is the key word in effecting a successful result. Carefully constructed, continuous-panel themes on either walls or floors can unite the various surfaces, providing a visual link throughout. Although the selected woods will determine the essential character, it is the grand effect that will carry the illusion. The tonal value of the woods is more important than subtle contrasts, and panel dimensions more important than detailed figuring.

Flow areas tend to be narrow and poorly lit and will benefit from the use of light-toned, reflective woods on wall surfaces, especially those at eye level. Distinctive, contrasting figuring tends to be strident and should be avoided unless the areas to be grained are generously proportioned. Continuous, horizontal accents, such as dado and picture rails, are particularly useful in implying movement. Where stairs are encountered, continuation of these lines in the form of the diagonal handrail will visually link the ground and first floors. These accents may be strengthened by creating panelled areas within their borders. Below dado height, panel systems can be darker and enclosed within strong borders. Above dado height, finishes should be pale and continuous with patterned woods displaying horizontal accents in, say, the form of mottled figures, such as fiddle back or block mottle. Natural breaks should be featured by dislocations in the grain figure rather than by any distinctive border.

Floor panel themes are very useful in visually indicating flow. Individual panel increments should show a sympathetic relationship of scale to the total area. Diagonal accents are preferable to rectilinear ones as their form implies a greater degree of movement.

Featured panel repeats provide schemes that require less commitment. These could be a series of simple parallel lines linking the various rooms. Alternatively, a linear progression of square or circular panels would be powerful enough to direct the eye. The possibilities of creating continuity themes within these areas is endless.

Left: *When presented with provocative architectural features, there is a great temptation to overstatement. In this instance, Peter Farlow has shown considerable restraint in graining only the spindles, central support and bannister rails. The essential structure and function of the spiral staircase are clearly stated, and visual continuity between the two floors has been established by continuing the technique along the gallery.*

Above: *Whether by accident or good design, the riser rebates in the central support, and shows more than an oblique reference to the structure of natural bamboo.*

The living room

The living room is a well-used, often informal area of the home. Its style tends to reflect the activities and aspirations of all those using it rather than any one individual, and the decor is usually dominated by objects that collectively reinforce the family unit. Silver trophies, groups of photographs and model aeroplanes may all be part of the essential character of the room. In this respect, wall surfaces should be presented as a neutral backdrop against which this reassuring clutter can be displayed.

   Although flat-paint surfaces are the obvious choice, unobtrusive woodgraining can complement the informality, warmth and intimacy of the room. Much will depend on the predominant colours in the existing scheme, but pale woods such as light oak or ash will provide a non-insistent backdrop. Tonal contrasts within the technique should be understated and 'woodiness' should be implied more in the coarse texture of the surface rather than by any distinctive figuring pattern. Areas at eye level should be enclosed within subtle panel definitions; those below waist height can be more clearly stated and, being less visible, they can contain more explicit graining.

Above: *Large rooms can accommodate a variety of textures, colours and surfaces, provided the disparate elements are featured against a unifying, and largely neutral, backdrop. Here, a comprehensive panelling system, grained in medium oak, provides an admirable background for the large collection of artefacts assembled in this traditional interior.*
Right: *Although visually distinctive, the mellow tones and mild character of pine provide a particularly successful backdrop to this unusual collection of objects – the warmth of the 'wood' making a striking contrast with the coolness of the marble bust.*

Where the atmosphere is more sophisticated, light, silky woods with regular mottle patterns or soft, stripe figures will provide a cool, more refined contrast.

If there are fewer family constraints, the living room becomes much more an extension of individual personality. Wood is capable of projecting many characters and a comprehensive woodgrained panel theme, on walls or floors, remains a practicable and cost-effective way of establishing style and atmosphere. While the natural limitations of the room and its contents are paramount, with sensitivity and planning you can construct themes that will benefit all situations.

For a scheme that requires minimal commitment, consider graining any of the wooden surfaces generally found in the room. These include the skirtings,

dado and picture rails and the window, door and picture frames. Although these areas are relatively small, they possess great strength of line. Where one technique is continued throughout, and where that technique affords a significant contrast between the graining and the background, the effect can be very powerful. Not only does it dramatically reinforce the geometry of the room, it implies a reduction in space, which can become visually distracting in small rooms.

Although these areas generally require the minimum of preparation, they are notoriously difficult to grain. Not only is the figuring complicated but there is also a tendency to overstatement. At a practical level, much time is spent up a ladder or on your knees, and it is well worth considering the implications and alternatives before beginning.

The cramped feeling associated with low ceilings has been visually relieved by constructing vertical accents in the panelled areas. Although warm, advancing colours arc difficult to orchestrate in confined spaces, Peter Farlow has producd grained areas that are sufficiently light and reflective (see detail below) to produce a cosy ambience, rather than one of enforced intimacy. Comfy furniture, plants and faux naive *art complete the cottage atmosphere.*

Doors and door frames are popular choices for graining. If the frame is grained using the same technique as the skirting, the visual continuation strengthens not only the presence of the door but also the geometry of the room by linking vertical and horizontal accents. A similar link can be made by graining picture and dado rails and window frames in similar techniques. You can treat the door itself in a conventional manner or create a fantasy in which panels, rails and stiles are represented in contrasting techniques and colours.

Fireplaces

Traditionally the fireplace is the focal point of the living room and generally reflects the period of the house or the stylistic preference of the owner. Although marbled fireplaces are becoming very popular, grained fire surrounds can be compelling. The overall tonal value should represent a significant contrast with the surrounding wall area and individual components of the piece should be grained using different techniques in order to strengthen their structure visually. Where the fireplace detail is uninspired, geometric grained areas can be constructed on the chimney breast to draw attention to the feature. These may be simple rectilinear column and lintel designs, more complex arched formations or panel systems.

A scaled-down, but sympathetic panelling system, grained by Peter Farlow, has been continued from the walls and echoed on the fire surround.

The bedroom

Although a bedroom is primarily used for sleeping, in many homes it doubles up as a study or peaceful retreat. Because it is essentially a private room, the decor can be more subjective and themes a little more exaggerated. For some, it might be the reflective qualities of tropical hardwoods that carry the illusion; for others, the essential feel of the room might be better served by the light-hearted pines or mild-mannered oaks.

Traditionally, bedroom furniture comprises bed, side tables, wardrobe and dressing table. Where furniture does not match, a single woodgrained theme can link the discordant pieces. When attempting this scheme, concentrate on the lighter colours of, say, maple, ash or satinwood. Rosewoods and mahoganies can become strident if the pieces are robust in nature.

Fitted melamine units provide a low-cost storage solution and are commonly found in modern bedrooms. These provide ideal surfaces for the woodgrainer. Inventive parquetry themes can enliven the dullest piece. Panel definitions should be identified as fine, distinctive contrasts in, say, the form of a gilded moulding enclosing detailed graining.

The bed tends to be the focal point of a bedroom and dictates the layout of the other furniture. Any painted features that confirm the arrangement will automatically become part of the natural scheme. A conservative effort may involve graining the bedhead and side tables in a matching theme. You can reinforce this by graining the area immediately above the bedhead in a contrasting technique. Focus features or a simple panel system can draw the eye to what is invariably the focal point of the room. Curved or arched systems, a popular format in medieval times, are enclosing and psychologically reassuring, and provide a pleasing contrast to strong horizontal and vertical accents.

Floor panelling systems are particularly effective in uncluttered rooms. Remember the bed takes up a lot of floor space so detailed graining is better concentrated in localized areas. These areas carry particular weight when they complement the organization of the furniture. Simple geometric areas of graining, placed at the sides and foot of the bed, are extremely effective.

Traditional features that include door, door and window frames, picture and dado rails and skirting board areas are useful in unifying the surfaces within the room. These elements contain strong vertical and horizontal accents, and while they may be useful in visually adjusting proportions they are easily overstated and tend to advance, producing a cage-like atmosphere.

Although you can grain entire wall surfaces, these are best treated as backdrops in the bedroom. Mottle patterns or simple toned stripes are preferred to distinctive figuring. Prominent rails and stiles have a functional quality that seems strangely out of place in a bedroom, and understated panel definitions are preferable. Butt joints indicated by discontinuity in the grained pattern, or fine pinstriped borders expressed as simple, contrasting outlines are good solutions.

Above: *In this realistic simulation, Peter Farlow has used contrasting mahogany patterns to identify the various rails, stiles and enclosed panels that comprise the window shutters.*
Right: *In this conventional treatment Peter Farlow has chosen a popular theme by graining window frame and surrounds, skirtings and architraves in a matching and continuous technique. Warm, dark colours have been selected, and the features stand out in sharp contrast to the wall surfaces. While in a small room the effect might have been overbearing, here generous proportions comfortably accommodate the exercise. And note how the grainer has used his skill to cleverly camouflage an incongruous, modern radiator sited in the window recess.*

The bathroom

The style and colour of the bathroom suite tend to generate the mood of the decor. Modern units are often self-coloured and usually feature a built-in bath with a free-standing toilet and basin. They are visually, and often spatially, dominant and the surrounding surfaces should be treated as backdrops rather than competing surfaces. As a word of caution, all techniques in the bathroom should be carefully sealed against moisture penetration.

The shapes of sanitaryware and their bright chrome fittings have little in common with the natural colour and form of wood, and so they present a challenge to the woodgrainer, especially if you want to develop a comprehensive theme. Colour, colour combinations and figuring contrasts are all important factors in resolving the problems associated with graining hi-tech finishes.

Traditional white enamel suites are less visually demanding, and graining treatments that express the warm character of wood can be used to counter-balance the clinically crisp nature of these surfaces. The yellow tones and mellow figuring of pine can be used to suggest a rustic simplicity, while mottle patterns or stripe figures project a more sophisticated appeal. Victorian-style suites featuring deep, free-standing baths and dulled metal fittings are more easily assimilated into the notion of wood as a practical, functional material. Coarse-grained woods, such as teak, medium oak or old pine, complement such schemes.

Standard treatments involve graining skirting and fitted surrounds in a matching technique. Horizontal surfaces, such as shelving systems or vanity units, can be developed as matching or contrasting features. These again you can strengthen by continuing the theme on to adjoining vertical surfaces.

Floor panel themes are warm underfoot and are particularly effective as a strong visual contrast to free-standing units. Where fitted units are used, continuing the pattern on to the bath surround will imply solidity. Echoing the theme on the wall areas enclosing the bath will strengthen the effect still further.

Although bathrooms tend to be small, wall surfaces that visually advance can imply a reassuring intimacy. While warm colours rather than definitive panel junctions should be used to achieve the effect, focus features in the form of simple book-end repeats can be used as charming contrasts to harsh, tiled surfaces.

In this Biedermeier-inspired bathroom, references to classicism abound. The 'wood' surround is a satin birch simulation, and the column and beam arrangement form a three-dimensional cage around the bathroom suite.

Above top: *Soft, subdued lighting is combined with the rich, warm colours and subtle sheen of the woodgrained doors, by Michael Synder, to create a cosy and inviting atmosphere in this washroom.*
Above: *Peter Farlow has applied mahogany graining to the bath surround, window shutters, achitrave, dado and picture rails to create a cohesive decorative scheme.*

The kitchen

Modern family kitchens tend to be the natural centre of the home. They are busy, functional areas dominated by bright packaging and noisy machines. Although fashionable, the cottage kitchen with its traditional Welsh dresser and colourful display of crockery is much less common than easy-care, hi-tech fitted storage units. However, if you hanker after a country look, a fitted kitchen can be transformed by an inspired painter, as shown below.

These system-built units, while undoubtedly practical, are soulless, uninspired pieces. However, their smooth, plastic surfaces present the perfect ground on which to paint. It is advisable to leave work surfaces and those wall areas at eye level free of distracting decoration. As for the remainder, it is a matter of how far you wish to take the technique.

Melamine surfaces are excellent for graining purposes. A permanent finish can be achieved by thorough preparation (see page 63). Once grained, you must seal the painted finish with at least two coats of polyurethane varnish (see page 69).

The type of wood you select will depend to a great degree on the character of the kitchen. But provided you retain large areas of light, flat colour you can introduce rich, dark-coloured, heavily figured woods, such as mahogany or teak. In order to soften the often regulated geometry of the units, try painting triangular panels or circular motifs in contrasting woods.

In addition to these units there are many other surfaces in the kitchen that are suitable for graining. While 'wooden' cookers, refrigerators and washing machines may be very much an acquired taste, results can be both provocative and humorous.

Far left: *The rustic simplicity, warmth and charm of natural pine has ensured its continuing popularity and use in many of today's modern kitchens. Although the dresser at first appears to be little more than a simple storage unit, close inspection reveals that many of the grained cupboards and doors feature elegant* trompe l'oeil *paintings, depicting a variety of fruit and crockery.*

Left: *Michael Synder created a fantasy limed oak effect for this cupboard door in the designer François Gille's kitchen.*

The dining room

Eating is more often than not a social activity – a time for conversation as well as food. Thus the dining room must serve as a meeting place, for family and friends alike, and provide a venue for functions ranging from the sophisticated dinner party, to the traditional Sunday lunch and the informal weekday supper.

To reconcile these various activities within one decorative scheme, subtle changes in lighting can be used to effect the necessary transitions in atmosphere. However, the difficulties of fitting a dining table and seating for six or more in the narrow confines of the average modern home places considerable emphasis on developing a greater visual sense of space in the room.

With this in mind, it is important to note that the wall areas above dado rail height will be the ones in prominent view. Thus, on any graining there, it is better to develop the textural, grainy quality of the wood, rather than any explicit figuring or distracting detail that will 'advance' the surface. Moreover, space and 'depth' may also be implied by enclosing grained wall panels within fine, and darker, contrasting borders. And arched and vaulted outlines will help to relieve the cramped feeling that can occur when many people gather in a restricted space.

Of course at dinner, attentions are mainly focused inwards, towards the dining table itself. This provides the woodgrainer with an ideal opportunity to demonstrate his or her skills to a 'captive' audience. Despite the fact that the table is usually cluttered with plates and serving dishes, a simple but detailed patterned repeat will prove to be the most effective – and can be extended to the dining chairs, if desired. Such designs, and they need be little more than a simple chequerboard pattern with surrounding border, are easily assimilated and interpreted by the brain – neither competing with their surroundings nor distracting the eye.

However, using the painted marquetry sequences suggested in the text, more complex and provocative patterns may be easily developed. Whether replicating specific woods or creating painted fantasies, strong contrasts in line and colour will seize the attention of the assembled guests, and provide a conversation piece. But do remember that the surface of a table is open to close inspection. So pay attention to detail.

In this spacious dining room, woodgraining has been employed in a comprehensive panelling system to produce an atmosphere of warmth and intimacy. Though the walls are visually advanced, the light ceiling and reflective quality of the grained surface prevent any sense of claustrophobia.

The study

'The study' implies a paraphernalia of desks, shelves and books. But more usually today a study is likely to be a retreat – a quiet, private place for contemplation and relaxation. In the modern family home the inclusion of a study must be regarded as something of a luxury. If it exists at all, often it is little more than a converted box room or child's bedroom. Space is at a premium and graining themes that relate to wall surfaces must be designed to reflect a light, airy quality. Tight satinwood pattern repeats, such as block mottle or raindrop figure, are formal but not strident. The mild tones of ash or maple are retiring and provide an opportunity to express linear growth ring figures that are both subtle and subdued.

Rectilinear panel definitions should be understated. Panel borders should be fine and expressed as a darker contrast to the enclosed graining. The dark panel surround advances and the enclosed graining retreats, producing a distancing effect. You can achieve a similar effect by graining the area below the dado rail height in a darker tone to that above.

Many traditional libraries demonstrate how the dark mahoganies and walnuts, especially when expressed as formal stripe figures, can be used to express an air of quiet, studious application. In smaller areas you can soften and lift their sombre tones by combining them with the livelier, reflective qualities of satinwood or maple. This format can be developed as panelling systems on walls or parquetry on furniture.

You can emphasize the studious nature of the room by focusing attention on the functional elements within it. The desk and writing chair, for example, can be treated as a central feature, with shelves, filing cabinets and storage units all united by a common graining theme. Rosewood, a classic cabinetmaking wood, is dark and formal; but lighter-coloured variations provide endless possibilities for a fantasy interpretation. And oak has particular associations with study – you only have to think back to your schooldays to remember robust oak desks and parquet floors.

Left: *Peter Farlow has used the mild colours and figuring of English oak to create an atmosphere of peace and serenity in this study.*
Above detail: *Simple dragging techniques define rails and stiles, while more elaborate figuring features in the enclosed panels.*
Right: *This post-modern bookcase has been grained in a satinwood fantasy. A variety of striped and mottled figures have been employed to define the various sections of the piece.*

Furniture and artefacts

Although any correctly prepared surface can be painted, there are two types of furniture that are inexpensive and can be substantially improved by graining — the ubiquitous melamine-faced fitted unit and the dark utilitarian furniture found in large quantities in secondhand shops. It is on such pieces that you can really begin to explore the versatility of paint and your ingenuity and skill as a decorative painter. The transformation of a nonentity into a desirable object is an interesting challenge, and a successful result will seem like magic.

Melamine furniture

Much of this type of furniture comes in kit form. This gives you a golden opportunity to grain and seal surfaces prior to assembly. Door panels are generally suspended on a frame and it is better to grain both doors and frame in the same wood for a more convincing effect. Continuing the technique on to the edges and inner surfaces of the piece can only reinforce the illusion. This ploy is particularly effective on larger pieces, such as full-length wardrobes.

The regimented, rectilinear form of the doors can be substantially relieved by carefully painted geometric designs, especially where right angles are broken up by circular or curved motifs.

Utilitarian furniture

Although this type of furniture tends to be poorly designed and made, painted decoration can lift a piece by redefining its visual proportions and drawing the eye to any promising detail. Once again, this is made possible by constructing simple panel systems and focus designs.

The functional aspects of furniture are reflected in their form and structure; identifying these elements in contrasting graining techniques will visually strengthen the piece. Severe contrasts produce fantasies, while subtle differentiation gives a more authentic feel. Decorating inside surfaces should give you a lot of satisfaction and delight the observer.

Another tactic when creating fantasy effects is to ignore totally the structural elements. Where a chest of drawers is to be grained, the piece can be treated as a solid cube, with figuring painted in continuous, flowing patterns over drawer and frame surfaces alike. Constructing the figuring as prominent diagonals rather than the normal rectilinear patterns can further reinforce its visual presence.

Graining different objects in matching techniques can help to unify furniture of varying origins. This effect produces a cohesive quality and is useful in linking the various disparate elements within a room.

Left: *A fine-detailed, mahogany-grained, French Empire mirror, complete with gold ormolu decoration.*
Above: *Grained boxes display, from the top: a framed oyster figure, tortoiseshell parquetry, and satinwood.*
Right: *This elegant, Regency style, open-shelved bookcase has been grained with a light-coloured, burr wood fantasy.*

CHAPTER FOUR

PANELS AND PANELLING SYSTEMS

Practical considerations, such as the size and type of tree, the quality of the timber and the problems associated with cutting, handling and transporting the material, impose certain limitations on the way wood is used as a cladding material. Traditionally it is displayed in the form of a simple geometric repeat, known as a panelling system.

As a decorative painter you can regard each panel as a canvas. The disposition of colour and the design of the various figuring patterns must produce a harmonious, balanced composition within the boundaries of each framed panel. Combined in the form of a panelling system, the sections must display a pleasing relationship of scale to one another and, collectively, to the proportions of the room. Expressing the woodgrained signature within this format can only enhance the credibility of the painted illusion ■

In this bathroom setting, Peter Farlow has grained bird's eye maple panels against a contrasting dark mahogany background. Although the mahogany is grainy and mottled, the lack of prominent figuring detail ensures that the eye is not distracted from the delicate bird's eye maple pattern. Simple mouldings surround the panels, and the rectilinear theme is echoed in the framed mirror and paintings.

Designing a wall panelling system

Although the graining of large areas may appear rather daunting, with careful planning, you can reduce the task to little more than an exercise in painting by numbers.

Measuring up

Having identified the wall area to be grained, it is necessary to take accurate measurements of the surface. Where a long span is to be tackled, use a tape measure rather than a short rule. Not all wall surfaces are perfectly true. If the area you are to tackle is square or rectangular, either measure the diagonals (these should be the same) or, alternatively, hang a plumb-line from the top corners to ensure that the sides of the section are vertical. Where the top or bottom edges are bounded by horizontal details, such as dado or picture rails or skirting boards, a spirit-level is useful in determining whether these are truly horizontal. When these dimensions are drawn up, any discrepancies can be disguised or accommodated in the painted borders.

Drawing up

Transfer the dimensions to graph paper using any convenient scale – say, 1 in to 1 ft (or 1 cm to 10 cm). Remember to include existing architectural details that may affect the composition. Make a mark at the intersection of the diagonals – this serves as the generating point when you are designing the theme and, later, transferring it to the wall surface. Overlay the graph paper with tracing paper and draw up your scheme. Bear in mind that the shape of the panels will fundamentally alter your spatial perception of the room and it is advisable to draw up several variations before making a final selection.

At this stage, it is as well to have a fairly clear idea in mind of the woods you want to simulate, their tonal values and their predominant figuring patterns. Colour approximations can be introduced on to the tracing paper in the form of crayons or watercolour to give a clearer idea of the overall effect. Indicate definitive figuring in pencil.

If the area to be grained is not quite true, construct a rectilinear scheme within the boundaries of the wall surface, leaving sufficient space for a surrounding border. Discrepancies can then be accommodated without disruption to the grained panels. Where more than one wall is to receive a panelling theme, you can draw up a more sophisticated diagram showing the total effect.

Transferring the design to the surface

The surface should be prepared as described on page

Where true verticals are to be constructed on wall surfaces, a plumb line is essential. A spirit level may be used to confirm the line you have constructed.

60. Using the intersection of the diagonals as a starting point, draught out your panel scheme on to the surface using blue chalk or a pencil. Again, a plumb-line and a spirit-level are useful in defining true vertical and horizontal lines. A long straight-edge or piece of string stretched between two fixed points provides a useful method of checking accuracy at every stage of the operation. Finally, indicate the areas on the wall that you intend to treat with particular techniques.

Graining the surface

Although the drying rate of the paint system you choose will influence the manner in which you tackle the surface, for all but the very experienced it is advisable to simulate different woods or different figuring patterns in separate operations, allowing adequate drying between the sequential procedures. Experienced grainers should be able to tackle two and often three simulations at the same time.

A chamfered straight-edge made from perspex approximately one metre long makes an ideal tool for creating clean panel definitions. You can accurately position the tool on the required edge and use a clean cloth to wipe off surplus glaze from its surface.

Before you begin graining ensure that you know the formulation of the coloured glazes you require and you have sufficient quantities of glaze to complete the operation in one go. Grain the panels and allow them to dry. They represent the largest area and any adjustments to the colour or tonal contrast of the surrounding borders that you have to make can be based on the totality of the theme.

Floor panelling systems

The graining of an entire floor surface is an operation not to be undertaken lightly. It causes disruption to the household and the room will be out of commission for several days. The justification for tackling such a scheme is that a wood-panelled floor is not only an eye-catching feature, it also generates an atmosphere of warmth and intimacy. Designs underfoot should be less strident, so graining may be far more automated and diagrammatic without sacrificing the authentic feel of wood. The design should be enclosed within a border that runs around the perimeter of the room.

Measure up and transfer the design as for walls. It is preferable to work on a demountable panel system (see page 56). When painting (and while the work dries), make every effort to minimize air currents as settling dust can be a problem. Take care, too, that you don't paint yourself into a corner!

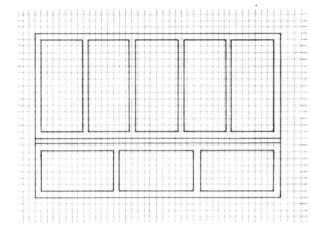

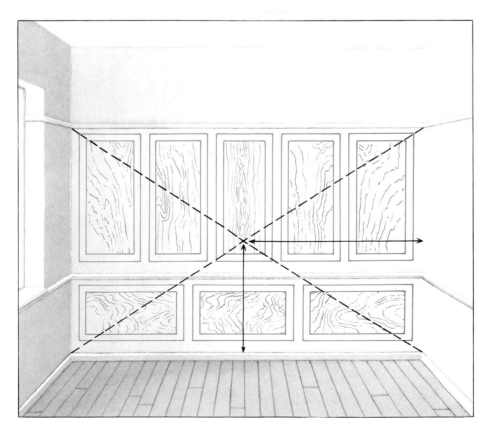

Top left: *The most effective way of designing a panel system is to first draw the design on graph paper.*
Left: *The simplest method of establishing the centre of a wall (or a floor, for that matter) is to stretch two pieces of string, diagonally, from corner to corner. Where the strings cross is the centre of the wall (or floor). This serves as the major reference point when transferring the design prepared on graph paper to the surface that is to be grained.*

Single panels or combination features?

Certain cuts of wood are so extraordinarily decorative that they are displayed as individual panels. The most definitive example of this is the 'feather figure' of mahogany. The feather can be as large as a metre square, in which case it may be used as a focal point on a wall. Although feathers may be enclosed within rectilinear, oval or circular frames, often the most successful solution is to display the feather against a contrasting, veneered backdrop with a minimum of distracting graining detail. Smaller versions can be featured individually on cupboard doors or table tops or positioned to form collective features. These features can be organized as simple repeats or used in more adventurous ways – as combination features, for example.

The natural material shows a progressive transformation of the figuring pattern through its girth. Where thin laminates are cut in the form of veneers, so slight are the variations in pattern that to all intents and purposes adjoining sections are exact replicas. By reversing one veneer, you have a mirror image. When these are placed adjacent to one another a 'book-end' pattern is produced. A mirror image of this combination is referred to as 'double book-ends' and produces a focus pattern. Once the principle is grasped the possibilities are endless.

Although the 'feather' example is a very distinctive feature, regular stripe, crown or interlocked grain figures can be used to form equally effective combination designs. Always be on the look-out for unusual applications of the technique. For example, where grain flow and knot formations on a piece of timber have been combined in the form of single book-ends to suggest the head of an animal.

Large-scale combination features used on wall surfaces provide yet another way of exploiting the illusion. Focus patterns narrow the vision and draw the observer into the surface; the feeling is one of enclosed intimacy. Conversely, simple progressions move the eye over the surface and in doing so suggest distance and space.

In this prestigious interior, the mahogany feather or plume has been used as a recurring motif on panels and doors. The continuity of theme, and the rigid symmetry of its expression, reinforces the formal atmosphere.

Panelling on furniture (marquetry)

Marquetry is merely a miniature panelling system suitable to the scale of furniture. The preparation is similar to that described for wall systems (see pages 52–3), with the exception that you can draw the design directly on to the piece. As a guide, coloured paper cutouts will give you an indication of the overall effect of the finished design.

The scale of the piece allows you the opportunity to give greater attention to detail, not only in the design but also in the graining technique. Remember, both will be subjected to closer scrutiny. Conversely, however, anonymous furniture may often benefit from explicit fantasies that take the eye away from the less-flattering aspects of the piece. In this respect, consider techniques such as graphic graining, colourful contrasts and distracting designs.

Remember to continue the graining over the edges of horizontal surfaces such as table tops to retain authenticity. And, where possible, lift the piece off the ground on to a raised work surface. This will make painting considerably less arduous and it will also keep dust intrusion to a minimum.

Marquetry became fashionable in the eighteenth century, and this beautiful table admirably demonstrates the decorative potential of combining a variety of woods into simple geometric shapes. The featured woods are: oak burr, satinwood, rosewood, mahogany, and zebrano.

Demountable systems

Demountable systems allow you to prepare grained sections off-site and then fix them into position in the room. Although this involves an initial outlay for material, there are substantial advantages to be gained, especially with the floor panelling system. Uneven surfaces are disguised and those that it would be impracticable to repair are covered up. Sound and heat insulation is improved as well. The system can be dismantled at any time and fixed in another room in the house when a change of decor is wanted.

The best material to use for a demountable system is hardboard. It comes in 8 × 4ft (2.4 × 1.2m) sheets in ⅛ in (3mm) or ¼ in (6mm) thicknesses. The smooth, hard surface will readily take oil-based or water-based paint and it is easily sawn to any required size and mounted. There is slight movement in the board, and sheets should be moistened on the back with water and stood for a day in the room where they are to be used prior to mounting.

When sawing hardboard it should be well supported at the cut edge to prevent tearing. Suppliers, though, will normally cut the material to size for an additional charge of approximately 10 per cent.

Where possible, graining should be carried out on the largest piece that it is practicable to use. This simplifies mounting procedures, reduces the risk of dimensional discrepancies that occur when the board is cut and minimizes the problems of graining over junctions.

Demountable wall panelling system

For this, use ⅛ in board. The design of the system is exactly the same as use for on-site graining, except that the size of the hardboard sheet should be given consideration when you are deciding on the size of the various panels in the system. This is not to say that you should be limited by the dimensions of the sheet, but you should try to ensure that hardboard and designed panel junctions coincide.

Care is required in translating the designed panel system to the hardboard sheets so that areas that are to carry the different graining techniques are clearly marked. Each sheet should be numbered on the back so that its position can be accurately registered when it is time to secure it to the wall. Where large areas are to be tackled, this type of systematic approach is vital.

Mounting the system Hardboard has little structural integrity and flat-headed nails or cupped washers are necessary to secure the board to the wall surface. Reference points indicating the position of the panels should be scribed on the wall prior to mounting. Panels should be suspended from the top corners and a plumb line, spirit level or set square employed to con-

firm a true vertical prior to final fixing. Where complex systems are to be prepared it is often better to locate and mount panels on the wall before they are grained off-site. This procedure removes much of the drama associated with any miscalculations made at the initial planning stage and it also ensures that the finished panels require a minimum amount of handling before mounting.

Demountable floor panelling system

The main aesthetic advantage of using hardboard sheets on floor surfaces is that the distracting 'long accent' of the floorboards does not interfere with the design theme. At a practical level, the sheet covers up unsightly surfaces or those in need of repair or refurbishment and it also reduces the presence of dust, noise transference and heat loss.

Use 8 × 4ft (2.4 ×1.2 m) ¼ in (6 mm) thick hardboard where the floor is reasonably sound and level. If a more substantial material is needed to accommodate irregularities in the underlying floor, use ¼ in chipboard. The surface of chipboard requires sealing with two coats of undercoat before it is ready to accept 'finish' paint.

Designing a demountable floor panelling system
Draw up the dimensions of the room on graph paper, overlay the plan with tracing paper and construct your design. The dimensions of the sheeting material must be considered when determining the size of grained panels. The butt joints of the hardboard should coincide with the panel junctions of the painted design. The design should be enclosed within a border that runs around the perimeter of the room. As well as being aesthetically pleasing, the border acts as a device for absorbing any irregularities in the room.

In order to determine the most economic use of materials and any off-cuts, you can overlay scaled-down cardboard cut-outs of your hardboard sheets on the design drawing.

Mounting the system Where floor areas receive a good deal of traffic it is important to secure the board at regular intervals – say, 1 ft 6 in (½ m) centres. Screws or nails should be countersunk, levelled with filler and touched up in a matching colour. Positioning and referring the cut boards in the room prior to graining will avoid problems associated with miscalculations or oversights.

Before mounting the system, the design should be drawn up on the floor to ensure the accurate positioning of the individual panels. Where the board does not extend to the skirting, chamfered wood sections or beading must be butted to the outer perimeter of the panels to prevent lifting and scuffing. This is particularly important at door entrances. Where curved features are designed as a focal point within a room, a proprietary filler should be used to chamfer the edge.

Furniture and artefacts

Similar demountable systems can be used on furniture or objects. Any material can be used provided it can be cut to a clean edge. Hardboard will be useful where a piece has become particularly shabby, and stripping and preparation is impracticable. Formica off-cuts can be bought extremely cheaply and make an excellent surface to grain. You can cut them into panel sections and glue them to the surface with contact adhesive after graining.

Less-substantial, plastic-faced materials, such as Fablon, provide ideal surfaces on which to test the overall effect of a panelling theme. Their surfaces require no preparation, they will take oil-based or water-based paint and you can easily cut them using a pair of scissors or a craft knife and steel rule. They are particularly useful in this respect for examining the potential of curved or convoluted designs. Such materials usually have an adhesive backing. After graining, you must fix them to the required surface with great care; there is no room for error as there are no second chances.

Peter Farlow grained the walls of this bathroom in a pine board simulation, completing the rails and stiles in a darker finish to suggest a simple panel system. Trompe l'oeil *wellingtons and a catch net complete this scheme.*

CHAPTER FIVE

PREPARING SURFACES

The ideal surface on which to grain should be perfectly smooth, hard and flat (see page 62). In reality, no matter how diligent your preparation, many of the surfaces that you will work on will fall far short of this standard. You must, therefore, apply a deal of common sense not only in selecting and preparing surfaces but also deciding on what technique is most suitable.

The organic nature of the wood is such that we understand and tolerate imperfections in the material. In this respect, while the simulation of hardwood veneers may require smooth, flat surfaces, varieties such as pine are less critical and surfaces that are somewhat less than perfect can be used.

Imperfections may be disguised by incorporating them as painted features, such as knots, tight-grain formations or panel junctions, within the design. Tight but irregular pattern repeats are particularly effective in disguising surface faults and the use of matt- or satin-textured paints or varnishes will minimize the reflection of underlying faults. As a qualification, remember that there is no excuse for poor preparation; paint has no structural integrity; it can neither bridge gaps nor fill cracks. Also bear in mind that glazes are insubstantial in themselves and any underlying unevenness will prevent the smooth and controlled distressing of the paint surface ■

Many surfaces can be grained with the appropriate preparation. Peter Farlow executed this limed oak graining over interior woodwork and a metal radiator.

Walls

Wall surfaces, especially those at head height (ie eye level), are likely to be scrutinized carefully and should be diligently prepared. Wall surfaces adjacent to a light source will show any imperfections, especially if the surface has a reflective sheen. In some cases, small imperfections or dips may be incorporated as details in the graining or repaired but large, undulating ripples, especially those in a vertical direction, can be extremely distracting and should always be replastered. A 6 ft (2 m) straight-edge laid along the surface of the wall in a vertical and a horizontal direction will reveal any major irregularities.

Plastered surfaces

Large areas of new plaster should be allowed to dry out thoroughly. They are slightly porous and should be sealed with size or emulsion before wallpaper is applied or, when they are to be painted, they should be lightly sanded first with a fine-grade paper to remove nibbs and then primed with some thinned emulsion or latex paint.

A proprietary fine filler (spackle) should be used for local repairs. Large cracks should first be agitated to remove loose plaster and then their edges chamfered with a scraper and the surrounding area rubbed down with coarse-grade sandpaper. A thick paste of filler should then be used to fill the crack and excess removed with a damp cloth to leave the repair slightly hollow. Once this is dry, a freer mix is then skimmed over the entire sanded area with a straight-edged spreader. Alternatively, a proprietary fine-surface filler may be used. The edge of a ruler laid over the repair will tell how level the job is. After drying, the repair should be lightly sanded with a medium-grade paper wrapped around a flat rubbing block. The last stages involve removing dust and over-painting the repair with matt emulsion.

With local, shallow flaking, loose material should be removed with coarse-grade sandpaper and the surface sealed with a proprietary stabilizer. Filler in the form of a thin paste can then be painted on to the surface with an old brush. When dry, sand flat with fine-grade paper and prime the surface as before.

Painted plaster

Where painted surfaces are in good order, they require only wiping over with warm, soapy water containing a little disinfectant and then drying with a clean cloth. If necessary, an abrasive sponge may be used to thoroughly clean and key gloss or eggshell finishes on any kitchen wall surfaces that have become coated with grease.

Localized flaking paint surfaces should be rubbed down with medium-grade sandpaper to remove loose material. Paint edges standing proud should be feathered and the surface sealed, levelled with filler and primed as before. As a rule, flaking paint is caused by some fault in the underlying surface, such as damp or a chemical reaction. Unless the fault is totally superficial, it may be necessary to renew the plaster to prevent the fault recurring.

Where the fault is widespread, the surface may have to be stripped using either a proprietary paint stripper or an industrial sander (available to hire/rent from most DIY stores or hire shops). The process is time-consuming, messy and costly. A better solution is to remove loose material with medium-grade sandpaper, stabilize the surface with sealer, emulsion it and hang heavy-duty lining paper.

Wallpaper

Embossed papers or those with textured relief patterns are unsuitable for decorative painting and will need to be stripped. Vinyl papers are impervious to water and are unsuitable for the application of water-based paint. They may, however, be sealed with undercoat or eggshell oil-based colour.

If the paper is lifting from the wall, this generally indicates damp plaster and regluing the paper will not prevent the problem recurring. If, however, the paper is in good condition and shows no signs of lifting, then it requires only wiping over with warm, soapy water and disinfectant and drying with a clean cloth. Small imperfections can be dealt with by injecting wallpaper paste through a hole pierced in the bubble using a plastic syringe.

Wallpaper joints that stand out can be disguised as panel joins, or incorporated in a surrounding border.

Fungal discolorations that occur in poorly ventilated, damp rooms should be treated locally with a proprietary fungicide. Small areas that have lifted can be reglued or cut out with a blade and a new piece let in. Gouges can be filled but sanding should be avoided. Use a damp cloth to remove and level excess filler.

Heavily patterned paper should be entirely over-painted to obscure the design. Certain colours have a tendency to bleed through a finish and over a period of time they may become evident in the grained surface. Matt black emulsion is the best visual sealer.

Where wallpaper stripping is necessary, although proprietary stripping solutions are available, copious amounts of warm water, a clean scraper and elbow grease are really the most effective ingredients. If whole rooms are to be stripped, wallpaper stripping machines are available for hire from most do-it-yourself stores.

Floor surfaces

Floor surfaces are less critical than walls, relying more on the overall impression created by the design than on any detailed subtleties of graining. Thus, preparation can be more relaxed and more flamboyant techniques and contrasts used to carry the illusion.

Unpainted floorboards

For the majority of people unpainted floorboards will be the surface they are likely to encounter. The longitudinal accent of the boards will undermine the credibility of any theme that is based on authentic panelled areas. While it is possible to disguise the joints partially with flexible fillers (spackle), the movement of wood is unpredictable and heavy traffic over the area will inevitably reveal some joints. Proprietary wood or general-purpose flexible fillers are available with variable dispensing nozzles and trigger rams for easy application. Fillers should be applied after the floor has been levelled and sanded.

First check to see that the boards are secured and that there are no prominent nails or splinters. If the boards have warped it may be necessary to use a light industrial flat-bed sander. Again, you can hire this, along with suitable grades of sandpaper.

Small cracks or faults in the boards should be filled with a proprietary wood or general-purpose filler, as described opposite for filling walls. Larger faults should be rectified by insetting a piece of matching timber. Square off the damaged section and span the inset between joists for the most permanent result.

After levelling the boards, lightly sand the surface, vacuum up the dust and apply two coats of undercoat

Graining works well with other finishes: wood and stone have been effectively combined in this disarmingly simple geometric floor design.

or proprietary wood sealer, allowing sufficient time between coats for drying.

Painted floorboards

Painted boards in good order need only be lightly sanded, vacuumed and primed with undercoat. Any minor repairs should be carried out as described for unpainted boards.

Painted boards in poor condition should be secured and bed sanded. Where thick coats of varnish have been applied, it may be necessary to burn off the varnish with a blowlamp first (varnish quickly clogs the paper). This task is best left to professionals.

Concrete floors

Unless concrete floors have been prepared to take a finish, they are generally level but extremely rough. Where large areas are to be grained it is necessary to lay a screed. This is really a task for professionals. For smaller areas, though, levelling screeds can be used and finer work can be completed with general-purpose proprietary fillers. Once again, a straight-edge is invaluable in locating imperfections and checking the evenness of repairs. When a reasonably flat surface has been achieved, the entire area should be vacuumed and then sealed with linoleum paint or a similar robust elastic film.

Vinyl floor tiles

Vinyl tiles can be overpainted provided that they are well stuck down and carry little textural relief on their surface. Although tile junctions are butt jointed, when overpainted the underlying pattern is often evident, and so it is as well to design a system with dimensions that coincide with the underlying tile configuration.

Before painting, the tiles should be scrubbed with warm, soapy water and disinfectant and left to dry thoroughly. You can then overpaint the tiles with undercoat. Where a sealing or isolating surface is required, use linoleum paint – it has more body, is opaque and can be overpainted with undercoat or eggshell.

Ceramic tiles

Ceramic tiles should be securely fixed and in good condition. They must be cleaned with warm, soapy water and disinfectant and allowed to dry before they are overpainted with undercoat. Cracked tiles will have to be removed and the gap filled with a proprietary filler. Joints should be filled and levelled with tile grout to disguise the pattern. Once again, where possible the panel system dimensions should coincide with some multiple of the underlying tile pattern.

Artefacts and furniture

As a decorative painter, you may be called upon to grain all manner of objects. Provided their surfaces are sound and reasonably smooth, they should be prepared, sealed and overpainted in the usual manner.

Wood surfaces

Wooden furniture is perhaps the most popular with decorative painters. Items for treatment are chosen often because they are either anonymous or darkened with age, stain and varnish. As a general rule, it is best to overpaint, but not necessarily decorate, the entire piece. Contrasting flat-painted surfaces often provide the essential balance for decoratively painted surfaces.

Sound, painted surfaces in good condition should be dry sanded with medium-grade paper and then primed with undercoat. Where the surface is grimy, first wash it with warm, soapy water and disinfectant using a scouring sponge. Wet surfaces should be thoroughly dry before overpainting with undercoat. Bright-coloured surfaces will probably need visually sealing with matt black paint.

Varnished surfaces are often covered with a layer of protective wax. This layer must be removed before priming by rubbing down the surface with medium-grade wire (steel) wool soaked in methylated spirits (denatured alcohol). After wiping with a clean cloth moistened with white (mineral) spirit, prime the piece with two coats of undercoat.

Where surfaces are chipped or peeling, loose material should first be removed and any proud edges feathered back with sandpaper. Exposed wood should be primed with a sealer before undercoating. Fill gouges with a proprietary wood or general-purpose filler (spackle), level with sandpaper and locally prime with undercoat.

If the piece is in poor condition, you might have to strip the paint surface completely. Proprietary paint strippers are readily available and are particularly useful for removing paint from delicate furniture or intricate details and mouldings. Follow the instructions carefully and let the stripper do the work. Premature scraping at the softened paint surface will only damage the underlying surface. Stripped surfaces must be soaked in water to neutralize any caustic chemicals (lye) left in the grain. When completely dry, fill any gouges and lightly sand before priming with a primer/sealer.

Large objects can be stripped by immersion in a caustic bath. This is a job for a specialist firm. Once again it is important to neutralize chemical residues.

As an alternative to stripping, you can burn the paint off with a blowlamp. This should only be carried out on robust natural wood surfaces. Wood composites are unpredictable and prone to blistering and buckling at high temperatures. Unpainted wood composites, such as plywood, blockboard and hardboard, can be primed with a proprietary sealer/primer and then undercoated.

The optimum surface

Although both oil-based and water-based grounds can be used as surfaces on which to grain, the optimum finish is provided by oil-based eggshell paint. After preparing the surface or object, it should be given two coats of eggshell. After allowing 24 hours for the first coat to dry out, it should be dry sanded with medium-grade paper, the dust removed with a cloth moistened with white spirit and the surface allowed to dry once more. The second coat should then be applied and given 48 hours to harden off. The surface should then be abraded with 600-grit wet and dry paper using warm water and a little detergent. For a superior finish, repeat this task using finer, 1000-grit paper. Press the paper against the surface with your fingertips and sand with a circular motion. For a perfectly flat surface, wrap the paper around a flat-faced rubbing block (left). Contoured or moulded surfaces can be abraded with a flexible sanding sponge. Mop the surface dry with an absorbent paper towel. A hairdryer may be used to drive off the last traces of water.

Plastic

A good deal of modern furniture is manufactured from wood composites faced with a thin, but tough, plastic laminate such as melamine or Formica. These are ideal surfaces for painting and require only the minimum of preparation. Lightly roughen the surface with fine-grade wet and dry paper using warm, soapy water and disinfectant. Once dry, overpaint with oil-based under-coat, eggshell or emulsion prior to graining.

Thin plastic film, such as Fablon, can be decorated with oil- or water-based paints. However, the material has no structural integrity and may soften, blister and distort if large quantities of oil-based glaze are applied. It is a useful material on which to test colours and techniques. Store samples in a folder for reference.

Plastic objects on the whole can be lightly sanded and primed with undercoat. Some of the very early plastics may react slightly with oil-based paint, so it is advisable to test these first by applying white (mineral) spirit locally to the surface with a cloth.

Metal

Painted metal surfaces in good condition should be lightly abraded with wet and dry paper using warm, soapy water. Once dry, undercoat the surface.

Chipped paint should be feathered back to a smooth edge with medium-grade dry sandpaper. If bare metal is exposed, a proprietary metal primer should be applied and allowed to dry. Proceed as before with wet and dry but avoid disturbing the primed area. Substantially chipped or pitted material may be salvaged by sandblasting (contact a specialist company), followed by a dusting, then apply primer and undercoat.

Glass

Glass surfaces require only washing with warm, soapy water. After drying they can be primed with undercoat. A word of warning, though. Woodgraining a glass object can disguise the brittle nature of the underlying material.

Fibreglass

Fibreglass and injection-moulded plastics are commonly used in the manufacture of such objects as garden furniture, statues and columns. The quality of the surface varies, but on the whole they require only a light sanding and dusting before the application of undercoat. Tiny pinholes that are sometimes in evidence can be filled by stippling on a proprietary filler mixed to a fine paste. Smooth the surface with a damp cloth as the mix begins to set to avoid sanding.

The cohesive qualities of featured graining are clearly evident in this stylish Art Deco theme. Dining table, chair and column are all united in a satinwood fantasy – the chic ambience demanding an understated pattern.

CHAPTER SIX

PAINT, GLAZES AND VARNISHES

This chapter introduces the various media that can be used to produce the woodgrained illusions shown in the following section. Instructions for mixing your own glazes are given, and advice on choosing suitable paint or glazes is included. Once you know the ingredients for graining, study the section on basic techniques (see pages 74–81) in the following chapter before you attempt any of the woodgrain simulations that begin on page 82 ■

Oil-based, water-based and acrylic paints and glazes can all be used for woodgraining. And clear varnish is essential to seal and protect your work.

Paint and varnish

The basic formulation of paint has remained largely unchanged for over three thousand years. While the petrochemical and rubber industries have done much to improve the variety and performance of paint, the effect of colour is still produced by the reflection of light from minute particles of pigment suspended in a transparent medium of oil or water.

Each system carries its own signature. Water-based paints produce clean, matt finishes, while oil-based paints tend to be lush and textural. Although the visual aspect plays some small part in influencing our choice as to which system to use, it is the performance of the respective paints in their fluid state that is most important here.

In the simulation of the various woods, you will use techniques that exploit the relative advantages of each type of paint. For example, a typical simulation could involve the following stages:

1 The initial 'rubbing in' process in watercolour.
2 Major graining procedures in oils.
3 Overgraining in watercolour.
4 Varnishing with oil-based polyurethane varnish.

This process may seem complicated, but in practice it is simple and straightforward provided that you follow a few guidelines.

Availability

Although on the whole the needs of decorative painters are best met by specialist suppliers, you should not experience too much difficulty finding what you need over the counter at most well-stocked paint suppliers. The popular woods have been identified by manufacturers and ready-mixed tinted glazes and ground coats closely approximating to the natural colour of the polished timber are available in a twin-pack system.

The do-it-yourself market has taken this a step further and you can now buy kits that include ground coat, graining medium, varnish and a patented graining tool. While it would be easy to decry the predictability of signatures produced by these kits, they do offer a simple introduction to woodgraining for the novice.

Water-based paint

Domestic water-based paints are available as popular emulsions. They come in matt, mid-sheen or gloss textures in the form of matt vinyl emulsion (latex paint), satin or silk emulsion and vinyl enamel. You thin them with water, they have little smell and dry rapidly.

You can apply this type of paint to any prepared surface other than bare metal, which would rust. Previously glossed surfaces should be rubbed down with fine grade sandpaper to provide a key for top coats. The drying rate of water-based paints is quick. The more thinned the paint the faster it will dry. While this feature is good for ground coats and allows the speedy superimposition of one coat on another, where the wet paint surface is to be distressed the time element becomes a critical factor. So if you are going to attempt graining over large areas or you want to work much detail into the surface, you will be better off using the slower-drying oil-based paints. You can extend the drying rate by adding a teaspoonful of glycerine to one litre of thinned paint. Also, drying will be retarded by maintaining a low ambient temperature in the work area and by keeping the atmosphere humid.

Matt and mid-sheen emulsions may be used as undercoats for either oil-based or water-based paints. However, there are two basic drawbacks. When dry, an emulsioned surface is unsuitable for sanding, since irregularities such as dust are echoed in superimposed coats. It is also porous, which means that the drying time of superimposed water-based paints is accelerated and the distressing period reduced.

Artists' water-based colours Because decorative painting is on a smaller scale than conventional decorating, it is often far more practicable to use artists' colours in the preparation of glazes for graining mediums. They are made from quality pigments and will produce better results than their domestic counterparts.

Although it is possible to prepare colour by grinding pure powdered pigment in water, an easier method is to use a water-based concentrate in the form of a thin paste. Although relatively expensive, artists' gouache colours are preferred. They come in a comprehensive colour range in tube or jar form. Tubes are particularly useful because you can dispense small, measured quantities on to a palette.

Because these paints are purely decorative, they contain none of the polyvinyl acetates (PVA) or poly-acrylic resins that act as binders and sealers in domestic emulsions. Dried pigment on a finish is readily put back into solution by the addition of water so it is therefore necessary to protect the dried gouache surface with a transparent coat of varnish. Also, when gouache colour is applied to an oil-based ground, fixatives or binders should be added to ensure the adhesion of the pigment to the surface.

These binders, which may include stale beer, a mixture of vinegar and sugar and fuller's earth, are discussed later in more detail (see pages 70–1).

Being water-based, gouache colours can be readily added to emulsion paint to create larger quantities of 'one-off' tints or custom colours. Again, the home

Paints, tools and equipment

Although some of the equipment illustrated here is specific to woodgraining, and is readily available from decorating shops or specialist suppliers, many of the items are in common use within the home. Artists' oils and gouache, notably the 'earth colours' such as yellow ochre, lamp black, burnt and raw Sienna, and burnt and raw umber, are available from artists' suppliers.

Steel Combs

Colour swatches

Scouring pad

Plastic combs

Low-tack masking tape

Pencil eraser

Sandpaper

Patented graining tool

Wet-and-dry paper

...stic sheeting

Cotton buds (Q-tips)

Pencil

Ruler

Check roller

Scalpel

Craft knife

Tubes of artists' gouache

Straight edge

Scissors

Tubes of artists' oils

Paper towelling

Clean, lint-free rag

Fuller's earth powder

...nsparent oil ...mble) glaze

Polyurethane varnish

White spirit (Mineral spirits)

Palette

Dippers

decorating market has anticipated the current demand for custom colours by introducing a system that uses measured quantities of tinter added to a neat white paint. A chart indicates the results produced when progressively larger quantities are used.

The great advantage water-based paints have over their oil-based counterparts is the speed at which they dry. Oil-based finishes must be left for at least 24 hours to harden before you can overpaint them. Water-based paints, on the other hand, especially in the form of thin glazes, dry in a matter of minutes. In situations where distressing techniques are being carried out with special brushes or tools, water-based colours ensure a speedy result and they are ideal for the impatient grainer. But take care. Only tackle those areas that you can comfortably distress in the drying time available.

Oil-based paint

Oil-based paints are available for domestic use in matt, mid-sheen and gloss textures as undercoat, eggshell and gloss. You thin these to the required consistency by adding white (mineral) spirit, they are relatively slow drying and have a pungent odour.

In decorative painting you can disregard the gloss variety as being too obvious. But you can apply both the undercoat and eggshell to all prepared surfaces as a ground coat. To eliminate any surface imperfections and provide the necessary key for graining mediums, the surface when dry should be lightly abraded with fine-grade sandpaper. Where quality finishes are required, it is preferable to abrade the surface with wet and dry paper. The use of warm water and a little detergent will prevent the paper from clogging and, where flat surfaces are to be grained, sanding down with a flat rubbing block will eliminate ripples or nibbs in the paint film.

'Flat paint', a product favoured by the trade, offers a quality alternative to undercoat. Its formulation is based on coloured pigments, drying oil and alkyd resins and it dries to a perfectly flat, slightly chalky texture. When using flat paint it is necessary first to prime the surface with undercoat. It is once again thinned with white spirit.

The drying rate of oil-based paint may be accelerated by the addition of small amounts of liquid dryers to the thinned paint and retarded by the addition of boiled linseed oil. This degree of control over the drying rate permits a more relaxed approach where lengthy graining procedures are to be carried out. In this situation, oil-based paints are preferred to their more unpredictable, faster drying, water-based counterparts.

Artists' colours in oils Although, once again, colours

may be prepared by grinding pure powdered pigment with linseed oil and white spirit, a better alternative is to use artists' oils. These are high-quality pigments bound in linseed oil and they are available in tube or jar form as a fine paste. They are thinned with white spirit and the texture of their rich colours can be varied from semi-matt to gloss by increasing the percentage of linseed oil in the thinned mixture. They are entirely compatible with their domestic oil-based counterparts and may be used to tint quantities of undercoat, flat paint or eggshell in the production of special colours.

Although quality paints and pigments are expensive, graining mediums as well-thinned glazes are capable of covering large areas. The additional expense of using artists' colours is offset by the ease with which the paint can be economically and accurately dispensed on to the mixing palette.

Artists' acrylic paint

Acrylic paints are a fairly recent innovation. They consist of coloured pigment suspended in a poly-acrylic, water-soluble base. Although thinned with water, the acrylic system is exclusive, with transparent mediums, drying retarders and a special palette all being necessary to exploit the various finishes. Acrylics come in a comprehensive selection of colours and produce a range of finishes that corresponds with both conventional oil-based and water-based systems. When substantially thinned, they appear as water-colour; used neat they carry the lush, textural quality of artists' oil paint.

Acrylic gloss and matt-textured mediums are available (from artists' shops and specialist suppliers) to give body to progressively thinned colour, allowing an extraordinary degree of control over colour transparency. The system offers one great advantage over conventional watercolour. Once the paint has dried the finish is non-reversible – pigment cannot be put back into solution by the addition of water. This allows the superimposition of water-based colour without disturbing the underlying finish. Consequently, smudged finishes are avoided.

For our purposes, however, the system possesses two drawbacks. First, despite the fact that a drying retarder is available, the length of time that the glaze or graining colour will remain mobile on the panel is difficult to determine. If the graining colour begins to set before the technique is completed, the panel must be abandoned and a fresh start made. Secondly, the addition of water to the neat paint initiates the hardening process and thinned colour must be kept continually moist and mobile on the palette or it will solidify and be wasted.

Varnish

The painted illusion is little more than a thin film of colour and, where necessary, you must protect the surface from mechanical and chemical change by applying a transparent coat of varnish.

At a decorative level, the appearance of many surfaces will be enhanced by a coat of varnish. Dark, matt finishes with little colour definition invariably appear lifeless and dull. A reflective film of varnish disperses light and intensifies the chromatic intensity or hue of the underlying colours, producing an effect of richness and depth. The effect may be readily seen by wiping a damp cloth over a piece of natural wood. The dark grain structure is accentuated and stands out in sharp relief against the paler background. A similar effect can be observed when the surface of dark marble is polished to a high lustre.

As a protective shield, varnish is regarded by many artists as a mixing blessing. Unprotected, the painted surface is quickly degraded by natural forces such as ultraviolet light and humidity. The creation literally fades before our eyes. Protected by varnish, the brilliance and subtlety of colour and the clarity of image is often sacrificed beneath a sticky yellowing film. Despite technological advances it is still regarded as a necessary evil by those who want the totally flat quality of watercolour.

Fortunately for our purposes, in many cases yellowing does not present a real problem. Yellowing often accompanies the natural ageing process of wood and implies a mellowing of character.

There are three main types of varnish suitable for woodgrained surfaces: water-based emulsion glazes, polyurethane varnish, and shellac or white polish.

Water-based emulsion glazes are available in matt, mid-sheen and gloss textures. They are applied neat but offer only limited protection. They show good resistance to yellowing and you can apply them over oil-based or water-based colour, but they tend to dry slowly to a sticky-looking finish.

Polyurethane varnish is regarded as very much the all-purpose varnish — you can thin it with white (mineral) spirit, it is slow drying and it is tough, especially when the surface is protected with several coats. It is available in matt, mid-sheen and gloss textures and it can be applied over dry oil-based or water-based paint surfaces. It yellows slightly with age and can be tinted with artists' oil paints.

Sometimes known as white polish, shellac is a natural substance secreted by coccid insects. It is available commercially either as rough crystals or as a ready-to-use product. The crystals must be dissolved in methylated spirits (denatured alcohol) or ethanol and because of the short pot life only sufficient quantities for the job in hand should be prepared. Although shellac offers limited protection, being partially soluble in alcohol and water, it is extremely fast drying and transparent, and so is a favourite among decorative painters. Where protection against mechanical or chemical damage is required, a final coat of polyurethane varnish is usually applied.

General advice

1 Never mix oil-based with water-based paints when they are in a fluid state.

2 Only overpaint those surfaces that are hard and perfectly dry.

3 When choosing paint always buy the best you can afford. Although the market is competitive, there is much variation in the quality of paint and it is up to you to identify your preferred finish and performance.

4 Thixotropic, non-drip paints should not be thinned. While they may be used as ground coats, they are unsuitable for the preparation of glazes and washes.

5 Follow manufacturers' instructions with regard to thinning and drying times and application procedures.

It is important, especially in bathrooms and kitchens, to protect paint from moisture penetration by applying several coats of clear polyurethane varnish.

Glazes, washes and graining mediums

Glazes and washes lie at the very heart of decorative painting and a fundamental grasp of their preparation and use is essential before you progress to the various techniques.

When paint is diluted or thinned with solvent, a point is reached where it ceases to be opaque. The paint becomes transparent; it has lost its covering power and the underlying ground colour is partially revealed in a slightly modified tone. The degree of transparency will be affected by the addition of more solvent, and control over this factor produces an infinite variety and subtlety within the painted finish. In this form the paint is referred to as a glaze or wash, and is now for our specific purposes a graining medium.

The technique of woodgraining demands a rather special approach. Detail features such as grain and figuring require clean and clear definition and glazes must be capable of exhibiting what are referred to as 'stand-up' qualities — the ability to retain clarity of print or pattern in the distressed surface.

Definitions and preparations

There is much confusion surrounding the terminology used to discuss glazes and washes. This is partially due to the conflict that arises when describing the various techniques, preparations and marketed products. For our purposes, a glaze can be described as any oil-based or water-based paint that has been thinned with solvent to a point where, when painted on to a surface, it shows a degree of transparency. A wash is a more substantially thinned glaze. When painted out, the pigment is so dispersed that only a trace of colour is evident on the surface.

Scumble or scumble glaze is the name applied to oil-based products that can be purchased from do-it-yourself stores or specialist suppliers. They may be colourless, in which case they are referred to as transparent oil glazes; or they may be ready-tinted, in which case they are generally referred to as coloured scumble glaze.

Scumbling the ground is a process that involves the initial application of colour to a prepared panel. It may require the use of one or more oil- or water-based glazes and the objective is to produce a tonally variegated background on which more complex, superimposed features may be painted.

Water-based glazes and washes

Water-based glazes are prepared by thinning water-based paint with water. They dry in a matter of minutes to a clean, matt finish. The speedy drying rate ensures that a minimum amount of airborne dust is attracted to the surface.

Emulsion-based glazes Preparation of glazes from emulsion paint will depend to a large degree on the strength of the pigment and the quality of the paint. Remember that thixotropic, non-drip paints are unsuitable for our purposes. As a rough guide, one part paint may be thinned with between one to three parts water, depending on the degree of transparency that is required.

A water-based emulsion wash may be prepared by thinning the paint with anything between one part paint to between four to eight parts water.

In both cases, painting out the results on white card will give an indication of the transparency of colour.

Emulsion-based glazes can be applied to matt or mid-sheen oil-based or water-based grounds. Emulsion grounds have certain drawbacks. They are unsuitable for sanding and tend to echo imperfections in the surface. Also, they are porous and the further accelerated drying rate severely restricts the time available for working or distressing the applied graining colour.

Washes and thin glazes dry at such a rate that, on the whole, they are unsuitable for use over water-based grounds. Although oil-based paints provide superior grounds, it is necessary to take precautions before applying an emulsion glaze. The surface should be lightly sanded or abraded with wet and dry or with a fine-grade paper to provide a key for the glaze. The surface should then be thoroughly cleaned and degreased with a domestic detergent.

The addition of a few drops of liquid detergent to the thinned glaze will reduce the surface tension of the water and prevent the applied glaze from 'cissing', or coagulating, into a series of tiny droplets.

Artists' gouache glazes and washes Neat paint is introduced on to the palette and thoroughly blended with a small quantity of water until a creamy, uniform consistency is achieved. Water is then added to dilute the mixture to the required transparency. The thinned paint should be stirred at regular intervals to ensure the even distribution of pigment and the results tested on a white surface before applying to the work area.

Unlike emulsion, gouache contains none of the additives that bind the pigments to one another and to the surface. As a result, graining mediums based on gouache colour or pure pigments should be combined with a fixative or binder.

In the past, recipes have used skimmed milk, stale beer or a mixture of vinegar and sugar as mediums for carrying colour. Although the use of 'found' materials has much to do with the innovative nature of decorative painting, apart from a certain novelty value these traditional mediums are unpredictable and can be unpleasant to use.

A modern alternative is to use small quantities of thinned white emulsion to stabilize the glaze. The emulsion additive contains sufficient binders and fixatives to give the glaze the necessary stand-up qualities. There is a penalty, however, since the hue of the glaze will be slightly subdued by the white pigment.

Fuller's earth powder is the most practicable and effective method of stabilizing waterborne pigment. Small quantities of powder are blended with water and the solution is then added to the prepared glaze. Alternatively, you can sprinkle a pinch of fuller's earth powder on to the prepared ground (preferably one that has been lightly abraded with fine-grade wet and dry paper) before the application of the glaze or wash. The powder is rubbed gently into the surface with a clean, damp cloth or moistened kitchen towel using small, circular motions. Then wipe the surface clean. The applied glaze is stabilized when it combines with the small quantities of fuller's earth left on the surface.

The drying rate of these glazes is extremely fast and, as with neat paint, a few drops of glycerine can be introduced into the thinned glaze to prolong the time you have available to work the surface. Do not add too much, though, or the work will not dry.

Acrylic glazes and washes Glazes and washes are prepared by blending paint with water and the necessary stand-up qualities, transparencies and textures are achieved by the addition of special gel mediums. The system is very specific and for those new to acrylics, a beginner's pack is available from most good art suppliers.

Oil-based glazes

The major advantage of working in oils is that the slower and more controlled drying rate allows the grainer far more time to work the surface. It is, therefore, suitable for large and small areas alike. There is a drawback, however – for the inexperienced, overgraining may be necessary to detail the finish and, in this situation, you have to allow time for the underlying glaze to dry out thoroughly before you apply further colour.

In the past, oil-based glazes were referred to as scumbles (see opposite) and were prepared by the grainer as and when required. The first task was to prepare the 'magilp' – a mixture of boiled linseed oil, turpentine and liquid dryers blended to produce a transparent medium. Quantities sufficient for a few days' work would be prepared and stored in sealed containers ready for use. The required colour was then made by blending selected artists' oils or pure pigments with turpentine. Measured quantities of 'gilp' were then tinted to the required colour by stirring in the necessary amount of mixed paint. Varying the quantities of the ingredients provided a flexible means of controlling drying rates. Today, ready-made products are available that simplify this procedure.

Transparent oil glaze is the modern equivalent of the magilp. This is a honey-coloured viscous liquid, which, when painted out, becomes colourless. It can be used neat or thinned with white spirit and can be tinted with artists' oils or pure colour pigment.

Ready-tinted oil glazes are available that correspond to the grain colour of most popular woods. In most cases, the appropriate opaque ground colour is specified and accompanies the glaze as a kit of parts. The glazes may be tinted with artists' oils to achieve a custom colour and they dry to a matt finish in two to three hours. Once dry, overgrain in oil or water.

An alternative is simply to thin an oil-based undercoat or eggshell with white spirit in the ratio of between one part paint to one to three parts solvent. A refinement of this is to thin white undercoat or eggshell as above and then tint the glaze by the addition of oil-based colour(s) – either artists' oils or pure pigments that have been blended with white spirit. Because the mixture always contains small quantities of white pigment, the vitality of the created colour is somewhat muted.

For graining purposes the most practicable method of preparing oil-based glazes and washes is to use artists' oils. Using a small, stiff-bristle brush, such as a fitch, colours are combined with white spirit on the palette. Where body or stand-up qualities are required, neat transparent oil glaze is blended with the prepared colour.

Oil-based glazes and washes are best applied to oil-based grounds, such as undercoat, flat paint or eggshell. These hard, smooth, non-absorbent surfaces have a slight sheen and glazes may be moved easily over the surface with great control. Although water-based grounds, such as emulsion, are suitable, their use is regarded as something of a compromise. Their surfaces are unsuitable for sanding and the consequent irregularities tend to choke applied glazes.

Where quantities of glaze are required, it may be necessary to transfer paint that has been blended on the palette to a dipper or jar before adding further solvent. An old stiff brush is perhaps the best tool for blending and transferring colour. Glazes in this form should be regularly stirred to disperse the coloured pigments. They may be kept indefinitely in well-sealed containers.

Although oil-based and water-based surfaces can be overpainted in oil-based or water-based colour, they should never be mixed in the fluid state.

CHAPTER SEVEN

TECHNIQUES

This section introduces you to the various tools and techniques that you may used to organize patterns in the wet glaze. The techniques are dealt with in three sections: the general 'broken colour techniques', the more specific 'graining techniques' that represent the unique qualities of the woodgrain signature, and the specific examples of simulated woods.

The simulated woods shown in this chapter have been chosen on the basis that they demonstrate a wide range of colours, textures and figuring patterns. The simulations represent a literal interpretation of a typical sample in that the painted illusion will closely approximate to the natural material. Within each wood type there exist tremendous variations, and until you are familiar with the various nuances of colour and pattern, it is advisable to refer to an actual sample, photograph or prepared sketch when graining. This reference will also be of great assistance when deciding on the main compositional elements on the panel. To illustrate the inspirational qualities of the material, variations are included that show how fantasies may be constructed by employing unnatural colours.

Individual sequences have been designed to illustrate the many different ways that the various paints and techniques may be combined to achieve a given effect. By following the various steps you should develop an analytical, systematic approach that you can apply when simulating not only other woods but also fantasy variations. Remember, experimentation, ingenuity and imagination are very much at the heart of decorative painting ■

In this free and uninhibited example of graining, the fantasy finish draws equal inspiration from the figuring of both wood and marble.

The broken colour techniques

The respective names for broken colour finishes result from the techniques or tools that are used in their creation, and so they have become traditionally identified as sponging, ragging, combing, dragging, spattering and stippling. (Sponging and ragging are omitted here as they bear no direct relevance to woodgraining.) Broken colour merely describes a film of colour that is neither continuous nor evenly distributed over the surface of the panel.

Additive techniques are those which involve applying wet glaze to the surface, while subtractive techniques are those in which tools or brushes are used to remove or displace wet glaze on the surface.

Stippling

This technique can be additive or subtractive. Subtractively, a flat-faced brush is used to strike the surface of the wet glaze. The small increments of colour that are displaced or removed reveal the ground colour, producing a very fine and regular speckled appearance.

Over large areas, a specialist stippling brush is used to rhythmically strike the entire surface, producing a uniform finish of great subtlety. Smaller areas may be tackled with any flat-faced brush such as a jamb duster, softener or fitch.

Additively, the delicate nature of the technique enables gradual shading and colour gradation to be accomplished by the intermingling of fine flecks of colour. It is also used to roughly merge colours before softening and blending.

The technique has direct application to woodgraining in the preparation of the fine-grained textured surface. After the graining colour has been rubbed in, patterned and softened, a light stippling will suggest the fine pores that are to be found in many woods. The technique is easily controlled, and stippling may be carried out locally or all over the panel at any stage of the simulation.

Additively, the technique may be used to simulate more prominent pores that are evident as surface features. Flecks of colour carried on the ends of the bristles are sparingly stippled on to the dried surface with a small brush. Softened in the long direction of the panel, they resemble the more elongated cells found in the hardwoods.

Stippling is a convenient way of physically levelling out a glaze, and in this respect it is particularly useful in spreading newly applied paint, disguising brush marks and breaking up the signature of strident prints.

Spattering

An additive technique, spattering releases flecks of glaze in a random fashion on to a surface. The process is generally a controlled and delicate operation: flecks of glaze are released on to the surface by, say, drawing a thumb over the stiff bristles of a stencilling brush charged with glaze or, alternatively, shaking glaze from the bristles by striking the shaft of the brush against a hard object.

Where two or three contrasting

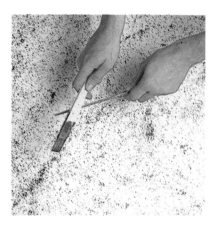

superimposed glazes are spattered on to a surface, a finish of great subtlety and delicacy may be achieved. Traditionally, the technique of spattering has been used in the simulation of polished stones such as granites and porphyries.

In woodgraining, the technique is of limited use. However, certain woods have irregularly spaced but clearly defined pores, and a spattered finish may sometimes be employed to produce a credible finish. Once again elongated pores may be produced by softening the spattered glaze along the grain axis of the timber.

Dragging

This technique is subtractive, and involves pulling a clean, dry brush over the surface of a wet glaze. The

bristles either displace or remove colour from the ground, producing a striped appearance. Where dragged finishes are required over large areas, consistency is the key word, and a systematic action, in which the brush is pulled in a continuous, unbroken stroke from the top to the bottom of the panel, is required. Although any flat, sharp-edged brush may be used, a specialist dragging brush will produce a superior finish.

The fine parallels produced by the technique have a general application to woodgraining. At the preliminary stages of a simulation, a dragging procedure known as brush-graining is often used to identify the course and flow of the prominent grain direction in the wet glaze. More defined parallels may then be applied using combs or brushes. At various stages in a simulation the surface may be dragged in a straight or wavy manner to reinforce the grain flow. The choice of brush will depend on the degree of subtlety that is required.

Combing

This is a subtractive technique in which a comb is pulled over the wet surface of a glaze. Colour is displaced or removed from the surface by the teeth of the comb to produce a series of parallel lines. Although sets of steel or plastic combs may be purchased at painting and decorating shops, home-made varieties considerably extend the flexibility and range of possibilities of the technique. Any semi-rigid material such as plastic or cardboard will suffice, and a few minutes' practice with a selection of combs will reveal the potential of this disarmingly simple, yet highly effective technique. Combs must be regularly cleaned with solvent to ensure a sharp print, and the ends of the teeth must be level with the surface.

The combed print is particularly strident, and for the purposes of woodgraining its diagrammatic print is reserved for fantasies.

In most timber, the growth ring figures of sap wood appear as a series of fine, roughly parallel lines. Combing procedures enable the painter to automate the creation of these regular figures. As a qualification it must be said that Mother Nature is neither repetitive nor predictable, and that a casual inspection of any piece of timber will reveal the irregular and variable nature of these parallels. For all but the most diagrammatic fantasies, much work is needed to soften the regimented print of the combed signature.

Combing is used at all stages of the simulation to reinforce the natural direction of the grain structure and create the discontinuous or serrated appearance of many of the prominent figuring patterns. A selected comb is pulled through the figure along the long axis of the panel. Combing may be continuous over the entire panel or localized.

Colourwashing

One of the most effective methods of subtly modifying the colour of a surface, colourwashing involves the application of transparent oil- or water-based washes to a surface.

Such washes dry fast, especially the water-based variety, so over large areas it is necessary to apply them quickly with a broad brush. It is difficult to keep a wet edge, and brushmarks remain an integral part of what is generally regarded as a rustic finish. It remains a good method of salvaging unsatisfactory finishes by visually cooling down or heating up the underlying colour. Progressive overlays will gradually deepen the tone.

For woodgraining purposes, the technique has two separate advantages. At a utilitarian level, washes provide an ideal method of gradually toning down strident finishes. A wash is prepared from the constituent colours present on the surface of the panel. When applied as a thin wash, the underlying colours are harmonized and their individual hues subdued. The technique of colourwashing is generally used as a final stage in the graining process to introduce tone and depth into the simulation. A wash slightly darker than the overall tonal value of the simulation is applied either locally or over the entire panel. The glaze is stippled and softened to create tonal variations in the required areas. Sections of dark glaze are then wiped out to reveal mottles and highlights. A gentle softening completes the panel.

Graining techniques

Isolating

Although the oil- and water-based systems should never be mixed in the fluid state, when dry, the finishes may be superimposed one on top of the other to draw the best characteristics from each. Aesthetically, the ethereal quality of water-based colour produces a clean, fresh quality not easily attainable in oil and it is often preferable where pale-coloured woods are to be imitated.

Where water-based glazes are used over freshly applied water-based gouache grounds, it is necessary to isolate or strap down the surface. Gouache colour is reversible, and unless extreme care is taken, the superimposition of a water-based glaze can result in the underlying pigment being returned to solution.

The process involves the application of a well-thinned coat of satin, mid-sheen polyurethane varnish. This will take several hours to dry. Before proceeding with the water-based glaze, lightly sand the varnish coat with fine wet-and-dry paper. Sprinkle a pinch of fuller's earth powder on to the surface, and rub in gently. Finally, wipe the surface clean with a dry kitchen towel before applying more glaze.

In some instances, heavy glazes may cause a certain amount of surface relief on the panel. To level out the surface apply several coats of white polish (a fast-drying varnish) – the varnish literally fills in the contours. When thoroughly dry it may be very lightly sanded using fine wet-and-dry paper and a flat rubbing block. Where oil-based glazes are to be overpainted the simulation may proceed, but if water-based glazes are to follow a layer of mid-sheen polyurethane varnish must first be applied.

Softening and blending

Although the technique has a general application throughout the whole graining procedure, it is of particular use in the creation of tonal variations and the softening of strident figuring. If we examine a piece of timber, we may see that the background colour is composed of many different tones of the predominant hue. This dappled appearance is a function of natural discolorations and the reflective quality of the surface.

Figuring is not a superficial quality of timber. Although the surface of wood is opaque, echoes of the prominent figuring patterns are discernible just beneath the surface. In the case of knots, the circular formation of the branch structure can be clearly seen receding into the body of the sample. As the features recede, the strength of line becomes blurred and their tonal value progressively muted. All these characteristics may be achieved by softening and blending procedures.

In the softening and blending technique the coloured pigments in a wet glaze are progressively feathered out or smudged by pulling a dry brush across the surface. The object is to produce a gradual tonal progression of colour from light to dark. Where different glazes are juxtaposed, softening and blending causes the pigments to intermingle, and produces a subtle gradation of colour one to the other.

Quality brushes ensure that brush marks are kept to a minimum. A hog's hair softener is generally used with oil glazes, and the finer badger softener with the more delicate water-based colours. These brushes are expensive, and the appropriate cleaning and storing procedures should always be followed after use.

Softening and blending with oil glazes Although it is possible to soften and blend all oil-based glazes, the technique is considerably improved by the presence of transparent oil glaze, which acts as a lubricating layer that provides enhanced mobility for the coloured pigments on the panel.

Where work on the panel is localized, the prepared glaze(s) should be blended with a small quantity of transparent oil glaze, taken up on a brush neat, and blended with the prepared glazes on the palette. The paint is then applied to the panel in the required manner, and softened using a hog's hair softener by gently flicking the surface of the wet glaze. A prominent accent can be suggested by concentrating the technique in one direction. The technique may be preceded by a light stippling technique to level the applied glaze.

Where softening and blending procedures are carried out over an entire panel, say in the preparation of a softly toned backdrop, it is sometimes desirable as a first stage to apply a layer of transparent oil glaze to the entire surface of the panel. This glaze may be used neat or slightly thinned with white (mineral) spirit, and is smeared over the surface using a clean cloth. The process is called 'oiling in the ground'. Glazes may then be applied and softened in the normal manner.

As a precaution, the blending brush should be cleaned with white (mineral) spirit at regular intervals to prevent the build-up of glaze in the bristles. The technique should continue only as long as the glazes remain mobile on the panel. In general, a working life of 15 to 30 minutes should be expected, depending on room conditions and the formulation of the various glazes.

Softening and blending using water-based glazes Water-based glazes and washes dry very quickly, and the length of time available for softening and blending is much reduced. A badger softener is capable of a finer finish than the hog's hair softener and is therefore preferable for the water-based work. The technique once again involves gently striking or stroking the surface of the wet glaze with the brush, applying just sufficient pressure to produce flexing in the bristles. Because of the fast drying rate of water-based gouache glazes, softening and blending should be carried out immediately the glaze is introduced on to the panel.

A few practice attempts will indicate the time that there is available. Although the oil-based system is preferable where much detail is to be worked into the surface, the water-based system is ideal for the initial rubbing-in and softening of the graining medium.

A few drops of glycerine stirred into the glaze prior to application on the panel will prolong the workable life of the glaze.

Artists' acrylic water-based paints have an advantage over gouache colour in that the system includes a retarder to slow down the drying rate and a transparent acrylic medium which bolsters up the body of the paint and allows pigments to be more readily dispersed and intermingled. As a qualification, it must be said that the workable life of glazes is rather difficult to determine, and it is all too easy to be caught out by a rapidly setting colour. Remember, once set, acrylic colour cannot be returned to solution by the addition of water, and so mistakes are more than a temporary set back as a fresh start must be made.

Wiping out

Wiping out involves the displacement or removal of wet glaze from the surface using any method or tool which is deemed suitable for the task. The choice of implement will be governed by the area to be tackled and the degree of detail required. The most versatile and useful tool is in fact the thumb nail. However, over large areas its use can become painful, and a substitute, known as a veining horn, is available from painting and decorating shops. This is used in combination with clean, absorbent lint-free cloth.

A pencil eraser, cork, card, cotton buds (Q-tips) or cloth are other useful tools. In preparing the objects, try to arrange for a clean edge that will completely lift or displace glaze from the surface. Removing stray glaze from previously wiped-out areas can be unrewarding and time-consuming.

There is a subtle difference between the effect produced when glaze is displaced rather than removed from the surface. Displaced glaze tends to build up at the sides of the wiped-out section, and appears as a prominent, slightly darker-lined edge. This hardening of line is often desirable in advancing the print or pattern. Conversely, the edge may be softened to imitate the gradual tonal transition that is evident in many growth ring figures.

The technique has two specific applications to woodgraining: the construction of complex growth ring figures and the preparation of mottles or highlights.

Firstly, figuring that is too complicated to be prepared by combing or brush-graining techniques must be either painted in with a fine brush or the pattern created by employing wiping out procedures. In both instances, the pattern will require further work to soften and detail the print. The end results are much the same, and the choice of technique is one of personal preference.

The wiped-out figure – which generally appears as a dark contrast to the lighter ground – is produced when glaze is wiped out to reveal the underlying colour. Because the process may be quite lengthy, the technique is usually carried out in oil paints.

The second important application of wiping out is known as mottling. Mottles appear as tonal highlights on the surface of wood. They may be grouped as regular patterns, or appear as isolated features, say around knots. Their shapes are varied and although specialist brushes known as mottlers may be used to create specific forms, materials such as leather or polythene wrapped around the finger are equally effective.

Mottles are created by wiping out areas of dark glaze to reveal a contrasting lighter-coloured background. The area is then gently softened to produce a gradual tonal gradation. The operation is relatively fast, and mottles are often introduced on to the panel at the preliminary stages immediately after the initial graining colour has been rubbed in.

Cissing

When neat solvent is introduced on to the surface of a wet glaze, pigment is dispersed from the point of application. Where the solvent is introduced in the form of a droplet, colour will retreat to form a circle; similarly a brush loaded with solvent and pulled over the surface will cause the pigment to retreat and a channel to be formed. Where glazes are composites of two or three hues, the solvent often causes the individual coloured constituents to separate out and arrange themselves in close parallels. It is very much a case of allowing the solvent to do the work. The skill of the painter is directed towards encouraging and controlling the happy accident. Removing excess solvent from the surface will bring cissing to an end. To avoid removing the patterned glaze, a light touch is required, and solvent should be soaked up with a tissue or cotton bud (Q-tip).

Solvents that cause oil-based colour to ciss are white (mineral) spirit, methylated spirit (denatured alcohol) or turpentine, while solvents that produce cissing in water-based colour are water, white spirit and methylated spirit.

Solvent may be introduced by any number of means. A stencil brush is perhaps the most practical tool for introducing small droplets. The solvent is taken up on the brush and released on to the surface by pulling a finger or thumb over the bristles. Larger, more random, amounts may be released by striking the shaft of a fitch loaded with solvent against a solid object. Individual droplets may be released from an artist's brush by just touching the surface of the wet glaze. Alternatively, absorbent fabrics or materials such as marine sponges or newspaper can be loaded with solvent and applied to the wet glaze in any number of forms to produce random patterns.

As a general application to woodgraining, cissing is usually carried out in conjunction with softening and blending procedures to produce tonal backdrops. The glaze is first cissed by transferring small quantities of solvent on to the surface using the bristles of the brush. After a few moments the glaze may be softened to produce all manner of tonal variations in the colour. While the technique is easy, there is much skill in exploiting its full potential. The strength of the glaze and the length of time that it has been allowed to set before the application of solvent will have a fundamental effect on the nature of the finish.

The technique has specific application to woodgraining in the preparation of burr-, pollard- and root-cut simulations. Solvent is released on to the surface in the form of droplets. The cissed glaze pools to produce ovoid and circular shapes. The characteristic dark eyes

of latent bud or branch formations that are a distinguishing feature of such cuts may then be spotted into their centres using a fine artist's brush. The swirling unrestrained grain formations of landscape or crotch figures may also be produced by forming cissed channels of spirit in the wet glaze.

Flogging

If you inspect the surface of most woods, you will see the grain structure as a series of pores. In the living timber these were the cells that were responsible for support or food distribution throughout the tree, and they vary in size depending on the particular species. Although they are primarily arranged along the longitudinal axis of the trunk, they may be evenly distributed or may locally group and follow wavy courses along the sample. This feature may be simulated by the process of flogging.

Flogging is a subtractive technique. A flogger is a long-bristled, long-handled brush that is used to literally flog the surface of the wet glaze. Colour is lifted or displaced on the surface by the tips of the dry brush to produce an overall textured appearance. The brush is generally used on its side, and the technique, a loose flicking wrist action, prevents the print from becoming mechanical or repetitive. The process is easily mastered, and with practice more complex, undulating formations and patterns may be produced. Although the brush should be cleaned with solvent at regular intervals to ensure clarity of print in the wet glaze, paint that is carried on the bristles may be redeposited on the surface during the flogging technique to produce small elongated pores. Although traditionally the flogger is the correct tool for the job, its print is often found to be too harsh. Where detailed, small-scale work is

required, the short edge of a softening brush or jamb duster allows a finer, more controlled technique.

Although grain formation appears very much as a background pattern, the textural nature of the surface is reinforced at all stages of the simulation by the process of flogging. After the initial rubbing-in of the graining colour, the surface is usually, stippled, softened and then flogged. And after the major figuring patterns have been introduced on to the panel, a light flogging is undertaken to break up the continuity of line. As a final touch, the overgraining glaze may be lightly flogged to strengthen more prominent grainy areas.

Check rolling

If you examine a sample of hardwood – oak, for example – dark, prominent pores may be seen on the surface. They are arranged along the long axis of the grain, and appear as an irregular series of elongated lines that are unevenly distributed over the surface. They are textural in nature, and their simulation may be automated by the technique of check rolling.

The technique is additive, and involves the use of a tool known as a check roller. This consists of an axle that carries several discs which are capable of rotating independently of one another. The discs have uneven, serrated teeth which carry the glaze. Colour is loaded on to the teeth by pushing the rolling tool through a shallow trough containing the colour. After testing the print on paper, the roller is moved over the panel to achieve the required disposition of pores.

Alternatively, a more controlled technique is to apply colour to the rollers with a brush while the tool is travelling over the surface. This allows a more random disposition of pores, and avoids a regimented, predictable print. The check roller may be used in the preliminary stages of a simulation, in which case the print is generally softened or fractured with combs; or in the final stages as a finishing touch to the overgraining procedures. (For a further method of introducing these pores see *Spattering*, page 74).

Mottling

The natural reflective quality of wood in which the surface is seen to sport highlights, mottles occur where the grain formation produces variations in the tonal value of the wood. Mottles may be found in isolated spots, say around knots, or arranged into regular and distinctive patterns that cover the entire surface. The simulation of mottles is referred to as the creation of 'lights and darks'.

Mottling is a subtractive process in which areas of darker glaze are removed from the panel to reveal the lighter background colour. Mottling may be carried out on the initial rubbed-in graining medium, after the combing of the graining medium or after overgraining.

Although the process may be achieved by combining wiping-out and softening and blending techniques, specialist mottling brushes have traditionally been used to automate the process. Mottlers are thin brushes with fine, soft bristles. The absence of a handle allows accurate control over the pressure on the surface and direction of attack. There are many varieties on the market, but as with all techniques, experience will determine the tool most suited to the task. The object is to remove and displace glaze cleanly from the surface with a light but positive pouncing and dragging motion. The displaced glaze will be formed into the familiar, fine-compressed grained parallels that initiate the reflective quality of the mottle in the natural material. The process is commonly referred to as 'flirting in highlights'. Gentle softening and blending techniques may be used to blur the print. Brushes must be cleaned at regular intervals with solvent to prevent a build-up of the glaze.

Parallel figuring techniques

The figuring of the various woods is concerned primarily with the creation of parallel patterns, and a variety of tools and techniques are available to assist in the automation of this procedure. It must be emphasized that nature is neither predictable nor repetitive. Where complex, convoluted figuring is to be faithfully represented, there is really no substitute for the diligent use of an artist's pencil or fitch.

However, stylised fantasies that rely on the diagrammatic prints of the various tools may be equally as powerful as their more photo-replicative counterparts.

Although the technique of dragging has already been examined in the section dealing with the broken colour techniques, it is necessary to cite a specific example to demonstrate the flexible manner in which techniques and tools may be used.

The initial rubbing-in of the graining colour involves two processes. First, the application of the glaze to the panel. And second the construction of a rough pattern that will determine the essential grain flow and compositional nature of the panel. The best tool for both stages is a poor quality small to medium brush. In the first stage, the uneven bristles allow the necessary irregular distribution of glaze over the panel. In the second stage the pattern is produced by the technique of brush-graining. The brush is pushed as well as pulled through the already broken glaze. The bristles flex and splay and produce enormous variation in the size, tone and direction of the parallel prints.

Combs For technique see page 75. In order to retain clarity of print, the teeth must be kept free of glaze by cleaning with solvent. While their use in the construction of regular sap wood signatures is invaluable, their use for more complex figures is limited.

The dragging brush For dragging technique see page 74. Although slight distortions and undulations may be produced by varying the pressure on the brush head as it is dragged through the glaze, the print remains regular and unnatural.

Again, old or poor-quality brushes with uneven bristles often prove most satisfactory. Wallpaper-hanging brushes are particularly useful (see page 80). The grainer's fingers may be threaded through the head of the brush, causing the bristles to reform into variable sized clumps as it is dragged through the glaze. Adjusting the angle of attack and varying the finger positions will all assist in producing the necessary variation in print.

Overgrainer The pencil overgrainer is used additively; glaze is taken up on the individual heads of the constituent brushes, and colour applied to the surface in continuous lines. Angling the brush and varying the pressure on the bristles will produce a variety of parallel signatures. It is essentially a brush for creating graining or overgraining detail on the dry grained surface; the absence of a handle allows great control of the brush.

The overgrainer is a less precise, but more versatile, brush than the pencil overgrainer. This brush is moistened with solvent and a selected steel comb run through the head of the brush to space out the bristles into individual brush heads. Once again glaze is taken up on the brush and applied to the surface in a continuous motion in order to create the required parallels on the graining surface.

Fan fitch The limitations of the comb become evident when it is necessary to simulate the irregular patterns found in flame or burr figures. This brush is capable of producing a wide range of roughly undulating variable parallels by angling the brush and varying the pressure on the bristles. A good deal of practice and experimentation is necessary to draw the full potential from the tool. Although it is essentially a brush designed for adding paint, interesting burr prints may also be obtained by taking up solvent on the fan fitch brush and cissing the wet glaze (see pages 77–8) in sweeping, curving movements.

Goose feathers An alternative to conventional brushes, the feather is quite difficult to control but the variety of parallel prints that may be achieved with this simple tool is well worth investigation. The feather is first cleaned in solvent, dried and then the fronds broken into sections with the thumb nail. Each section may be regarded as an individual brush head. A pair of scissors may be necessary to clean up and level the edges of the feather. Glaze is taken up on the feather and applied to the panel in a continuous motion. By angling the feather, the distance between the parallels may be varied in much the same way as the fan fitch, and in this respect it may be used to create crown or flame figures. Because only limited amounts of glaze may be taken up on the fronds, continuity of line may be lost if long uninterrupted strokes are made. As an alternative, solvent may be taken up on the fronds and the feather used to ciss channels in the wet glaze.

The unpredictable nature of the feathered signature means that it is more suitable for use in the preparation of instinctive fantasy woodgrains, rather than detailed replications.

Patented graining tools Tools are available that are capable of representing the tight, continuous parallels and arcing parabolas of the flame or crown figure. Although the prints are limited and slightly predictable, they are well worth investigation, especially where large schemes are to be undertaken. The signature is produced by wiping out areas of wet glaze. As the tool is pushed or pulled through the glaze in a continuous motion, the wrist is gradually rolled to bring different areas in contact with the surface of the panel. Graining tools may be purchased as an individual item, or included as a graining tool in a kit of parts that includes graining mediums and instructions.

The techniques for using one of these tools – the check roller – is described on pages 78–9. The roller itself is shown on page 67, along with an alternative graining tool.

THE MODIFIED WALLPAPER HANGING BRUSH

The woodgrainer should always be on the lookout for tools or brushes which can be modified to assist in the production of a particular print.

The wallpaper hanging brush can be used as a dragger, and as such has three advantages: its width, normally between 6 and 9 inches (15–22.5 cm), which allows large, continuous sweeps to be made through the wet glaze; the way that the poor-quality bristles splay, *thereby introducing a degree of unpredictable, and therefore 'natural', variation in the print, and finally, the fact that the brush is readily available and relatively cheap.*

Note: Further variation in the print can be achieved by varying the pressure on the head of the brush, causing the bristles to splay, and adjusting the position of the fingers in the bristles.

1 *Normally the brush contains three rows of bristles. Use a craft knife, or scissors, to cut out two rows as they exit the brush head.*

2 *The remaining row is sparse, so you can feed the fingers of both hands through these bristles as the brush is pulled through the glaze.*

Brushes

There is a wide range of specialist brushes available for the different aspects of woodgraining. But in practice, many brushes can be employed for more than one purpose. For example, a softening brush can be used for dragging or stippling, and a fan badger could be used for detailed stippling or flogging work. However, if you choose to work in both oil and water-based mediums, you are advised to acquire a range for each system.

Flat fitches

Artists' sables

Badger softener (large)

Stencil brush

Hog's hair softener (large)

Flogger

Hog fan overgrainer

Badger fan overgrainer

Soft hair mottler

Varnishing brush (large)

Badger fan

Hog's hair pencil overgrainers

Overgrainer

Varnishing brush (small)

Badger softener (small)

Large bristle grainer

Medium flat brush

Small flat brush

Bristle mottler

Harris

MAHOGANY

In addition to its rich coloured tones and handsome figures mahogany shows remarkable stability after seasoning and is extremely durable, being resistant to both fungal and insect attack. The lightest mahoganies have great reflective qualities and yellow or pinkish hues, while the darker varieties exhibit deep red to purple-brown shades.

West Indian mahoganies (the first to arrive in Europe) are darker in colour and silkier in texture than Cuban, American or African varieties. They exhibit close straight-grain characteristics and are harder and heavier than any other genus. African mahoganies exhibit bright pink or salmon colours which gradually darken to light or medium reddish-tan with a slight purple cast.

In both varieties the erect axis of the tree is evident in the long straight-grain characteristics. While heart grain figures tend to be rather uninspired, the interlocked grain structure is responsible for the many and varied patterns which result from stripe and broken stripe figures. The long straight axis of the tree permits rotary sawing, and in many cases the rare and decorative blister figure is produced. Mahogany exhibits over 25 clearly identifiable figures; the simulations indicate some of the more graphic possibilities. In reality, over a typical sample of wood, the patterns of the various figures gradually change from one form to another.

In tonal value mahogany sits somewhere between the lighter satinwoods and the darker walnuts. Although the wood tends to be formally exhibited in a striped format, its mellow tones and varied figuring patterns offer a wide range of possibilities in graining wall surfaces and furniture. Extensive mottling lightens and lifts the wood to a point where it may be used as an anonymous backdrop. The definitive crotch and swirl figures may be lightened, darkened or adjusted in scale.

Stripe figure

1 The basic ground colour for mahogany, nectarine in an eggshell finish, tends to be rather bright, and you can make it more red or grey by adding small amounts of mid-red, black or white artist's oils. Mix a ground coat of your preferred shade and, using a standard decorators' brush, apply it to a previously prepared surface (see page 62). Allow to dry for 24 hours. Using artist's oils, prepare a mid- to deep-tone brown glaze by blending burnt umber with a little black and thin to transparency with white spirit. Apply glaze loosely to the whole panel.

2 Using a 2.5 cm (1 in) decorators' brush, immediately brush-grain the surface. Organise the wet glaze into a series of roughly parallel bands of tone by pulling the brush in a continuous motion from the top to the bottom of the panel. Vary the pressure on the bristles and the angle of the brush head to produce the necessary variations of tone. Darker stripes may be introduced by charging the brush with small quantities of deep brown oil colour.

3 Soften the striped pattern by gently striking the entire panel across the long axis of the stripes with a badger softening brush. You should also make occasional up-and-down strokes to confirm the striped format. Allow to dry for 24 hours.

4 Prepare a slightly deeper-toned glaze by blending burnt umber artist's oil paint with a greater percentage of black. Thin this with white spirit and use a 2.5 (1 in) decorators' brush to brush-grain the entire panel, following the long direction of the stripes.

5 Lightly flog the panel with a flogging brush to texture and lighten the surface. Using the narrow edge of the brush, work in vertical lines along the lengths of the individual stripes. Because the arrangement, density and size of pores in the wood are irregular, you should vary the technique by concentrating on some areas and leaving others plain.

6 After the completed panel has been left for 24 hours to dry out, apply two coats of mid-sheen or gloss polyurethane varnish.

Crotch figure

1 Using a standard decorators' brush, apply a nectarine eggshell ground coat onto a previously prepared surface (see page 62). Allow to dry for 24 hours.

2 Using artist's oils, prepare a mid-to deep-tone brown glaze by blending burnt umber with a little black. Thin to transparency with white spirit and a little transparent oil glaze. Brush-grain the panel in a series of sweeping continuous arcs using a 2.5 cm (1 in) decorators' brush. Work systematically from the outside of the panel towards the central spine.

3 Using artist's oil colours, prepare a deep brown-black glaze. Strengthen the ribbed pattern by introducing a series of darker stripes to the panel, with the 2.5 cm (1 in) decorators' brush. Stagger the stripes on alternate sides of the central spine. The colour should be more concentrated at the centre and thinner towards the edges of the panel.

4 Using the original rubbing-in brush, stipple the glaze prepared in Step 3 into the central area to create dark but broken irregular blocks of colour. This will strengthen the spine and give the effect of 'vertebrae'.

5 Using a modified wallpaper brush (see page 80), reinforce the rib and spine pattern by brush-graining the entire surface. Start at the base of the panel and move upwards, and work inwards from the edges of the panel towards the central spine, following the direction of the rib formation. Immediately before reaching the spine, pull the brush inwards and downwards towards the centre, using a sharp flicking motion. The italic quality of the flat brush produces a series of wispy undulating parallels. Systematically treating the arcs in this way produces a 'centralized V' on the panel. Avoid disturbing the darker spine. (The glaze will remain mobile for some time and you may have to return to Steps 2,3 and 4 before you achieve a satisfactory result.)

6 Use a small badger brush to soften the ribs. Work outwards and upwards, striking the surface of the

panel more or less at right-angles to the direction of the ribbed pattern. Then allow the panel to dry.

7 Prepare a black glaze using artist's oil colour. Take up this colour on a fan badger brush and stipple the surface of the panel with the side of the brush. The bristles are sparse and flexible and produce the fine wispy effect of the long pores. Work outwards from the centre, following the line of the ribs.

8 Soften the pores with a small badger brush. Work in the direction of the ribs. Define the pores more clearly in the lighter areas. Allow to dry for 24 hours.

9 Seal and protect the finished panel with two coats of mid-sheen or gloss polyurethane varnish.

Note: This simulation has been described using artist's oil colours throughout. As an alternative, you can carry out the initial stage of this technique, up to and including Step 6, in water colours. Where the colour in Steps 1 to 6 is oil-based, you can continue in either oils or water colours. If you begin with water colours, you must complete the panel in oils.

Although the sequence may appear long-winded, a few trial attempts will reveal the speed at which the crotch figure can be created. To save time, steps 7 and 8 (the introduction of fine pores onto the surface) may be omitted, if less detail is required. Fantasy variations, like those shown on the right, often take only a few minutes to produce.

Note that some of the rib formations remain incomplete as they broach the outer edge of the panel. When producing panel simulations, such features will add to the credibility of the simulation, as incomplete figures are common on cut panels of genuine wood.

Although the crotch figure is prized for its symmetry, it benefits compositionally from being displayed at a slight angle to the true vertical of the panel.

VARIATIONS

A In this version the base colour is matt silver, and the overgraining colours predominantly black with the addition of small quantities of yellow ochre to produce a slight green tint. The composition shows how the swirl begins to emerge in the top left-hand corner. The spine is quite dislocated and angled to produce a feeling of movement on the panel.

B Here the base colour is an acidic green with overgraining in shades of green and yellow. The structure of the grain is more orthodox than in variation A.

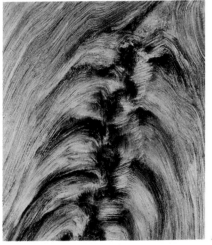

Blister figure

1 Onto a previously prepared surface (see page 62), apply an eggshell finish nectarine ground coat, using a standard decorators' brush. Allow to dry for 24 hours. Using artist's oils, prepare a rich brown glaze by blending burnt umber with a little black. Thin this to transparency with white spirit and a little transparent oil glaze. With a decorators' brush of 2.5 cm (1 in), brush-grain the panel to produce a series of fine undulating vertical bands. As you work, angle the head of the brush and keep changing the pressure on the bristles to give the necessary variations of tone.

2 Wipe out small areas of glaze with the sharp, long edge of a pencil eraser to produce a series of mottles across the entire panel. Displace the excess glaze onto the bottom edges of the mottles. Vary the angle, size and shape of the mottles by angling your wrist and adjusting the amount of the eraser's edge that comes in contact with the glaze.

3 Make a series of predominantly downward strokes with a broad badger softening brush. The areas of glaze along the lower edges of the mottles will smear, to give gentle tonal gradations and a softer effect across the whole panel.

4 Pull a pencil overgrainer through the glaze to texture the panel and give the effect of the grain formation. The technique involves pulling the brush from the top to the bottom of the panel in a continuous, hesitating, swerving motion. Work systematically over the panel, but avoid producing a regimented pattern – the curves made by the adjacent passes of the overgrainer should not be in sequence.

5 Wipe out more glaze, this time using the short, sharp end of the pencil eraser. With the corner of the eraser, concentrate on the elongated thread-like filaments that travel down the length of the panel, making them wind in and out of the mottles, sometimes interconnecting with adjacent strands.

6 Now soften the entire panel with a large badger brush. Do this by

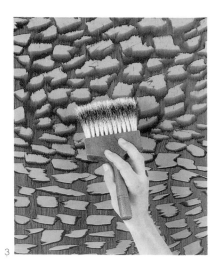

striking the glaze both down and across the surface. Again, it is important to vary the technique; you can make some areas very soft while keeping others quite sharp.

7 Using the thin edge of a flat badger brush, gently flog and texture the glaze, working along the vertical bands that are defined by the thin filaments. By this time the glaze may be quite tacky and the brush will redeposit some of the colour that it picks up from the surface, to leave a series of elongated black pores. This controlled flogging procedure completes the first stage and when you are satisfied with the effect, you should leave the panel to dry for 24 hours.

8 Although well textured, the tonal variations on the panel are a little strident. So prepare a dark wash by blending burnt umber and black artist's oils with copious amounts of white spirit. Use this colour to harmonise the tonal range of the panel. Using a 2.5 cm (1 in) decorators' brush, apply the wash unevenly to the surface and brush it out lightly down the whole length of the panel. Allow to dry for 24 hours.

9 Seal and protect with two coats of gloss or mid-sheen polyurethane varnish. Although the final composition shows a fairly even pattern, you can vary the scale and arrangement of the mottles to suit your own taste and the setting of your particular panel.

Note: For convenience this simulation was carried out in artist's oil colours throughout. But you can carry out Steps 1 to 7 in either artist's oils or water colours. If you use oils for this first stage, you can continue with Steps 8 and 9 in either oils or water colours. If you begin with water colours, you must finish in oils.

VARIATIONS

A Here a black eggshell base colour has been over-grained with gold artists' oil. The mottle patterns have been broken up into distinct sections, producing an abstract effect not dissimilar to the markings found on leopard skin.

B In this variation the base colour is white eggshell, and the overgraining colour black artists' oil. The mottled pattern has been arranged to produce a flowing quality in the composition. The edge of an eraser was used to wipe out the longitudinal filaments.

Fiddle back figure

1 Using a standard decorators' brush, apply an eggshell finish nectarine ground coat onto a previously prepared surface (see page 62). Allow it to dry for 24 hours. Using artist's oils, prepare a deep brown glaze by blending burnt umber with a little black. Thin this to transparency with white spirit and a little transparent oil glaze and brush-grain the panel with a 2.5 cm (1 in) decorators' brush to create a series of tight, roughly parallel, vertical, tonal bands.

2 Brush-grain a series of dark wavy lines into the glaze by manipulating some of the colour already on the panel into more concentrated bands. These wavy bands will serve as boundary lines for the rows of mottles that you will create on the panel in Step 3.

3 With the long clean edge of a pencil eraser, wipe out a series of fine horizontal mottles in the glaze. In each mottle, make sure that the excess colour is pushed down towards the bottom edge. Work systematically down the panel, using the dark undulating lines as confining boundaries. Slightly vary the angles of adjacent bands of mottles – as well as the shape of each individual mottle – so that the pattern does not look too regimented.

4 Using a large badger brush, start the process of softening by brushing lightly across the panel. Then work mainly in a downward direction to create a soft gradation of tones within the mottles. You should soften some areas completely but leave others quite sharp. Now leave the panel for 24 hours to dry out thoroughly.

5 Prepare a deeper-toned but more transparent oil-based glaze by blending burnt umber and black artist's oil colours with white spirit. Using the 2.5 cm (1 in) decorators' brush, apply the glaze to the entire panel in loose irregular patches of tone.

6 With the 2.5 cm (1 in) decorators' brush, brush-grain the wet glaze prepared in Step 5 into a series of fine, tonally varied stripes that

7

8

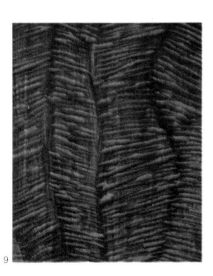

9

roughly echo the underlying pattern of mottled bands.

7 Use a small badger brush to strike the glaze across the length of the stripes, to soften the panel. Once again, you should vary your technique as you work across the board, so that you leave some areas of the graining well softened and others quite sharp.

8 Texture the surface by flogging the panel with the edge of a long-handled flogger. Work systematically down the bands of mottles, but do not flog the entire surface consistently – you can leave some areas quite lightly flogged. The process of flogging will displace some glaze and remove a certain amount of glaze altogether, so the areas where you concentrate on this technique will be generally lightened. Allow the panel to dry for 24 hours.

9 Seal and protect with two coats of gloss or mid-sheen polyurethane varnish. The finished panel shows the characteristic fractured appearance of fiddle back mahogany. You can adjust the scale, width and regularity of the pattern repeat to suit the specific area that you are going to grain.

Note: Fiddle back and other broken stripe figures (eg block mottle, raindrop and rope) are common to all tropical hardwoods that have interlocked grain characteristics (see page 16). Although featured panels in the step-by-step simulations show a more or less even distribution of pattern, in reality samples may exhibit three or four distinct figures, each gradually transmuting from one form to the next over the surface. On larger panels, you should incorporate the various figuring combinations in your design, rather than a single figure, for authenticity.

This simulation is carried out in artist's oil colours. It is possible to use water colours, but the graining process is quite lengthy and there is the possibility that water colours will dry out too quickly.

VARIATIONS

A The base colour and graining colours are the same as those used for satinwood simulations. The fine, horizontal fiddle back markings have been created by cissing the striped format with a pencil overgrainer charged with white spirit.

B In this dramatic fantasy variation, a viridian green and burnt umber glaze has been grained over a turquoise base. Plastic combs have been used to wipe out colour, and additional tension has been created in the composition by angling the pattern in adjacent sections.

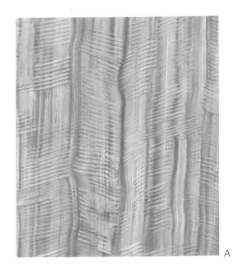

A

B

Swirl figure

1 Use a standard decorators' brush to apply an eggshell-finish nectarine ground coat to a previously prepared surface (see page 62). Allow to dry for 24 hours. Mix a mid- to deep-tone brown glaze by blending burnt umber artist's oils with a little black. Thin the glaze to transparency with white spirit and a little transparent oil glaze and use it to brush-grain the panel in a series of continuous, looping swirls and curls. Use a 2.5 cm (1 in) decorators' brush and work mainly diagonally across the board, rather than vertically. The parallels should show variation in tone and width, so adjust the angle of your brush and the pressure on the bristles as you work.

2 Use a modified wallpaper-hanging brush to brush-grain the panel in a series of continuous sweeps that roughly follow, but do not entirely match, the underlying pattern.

3 Soften outwards from the bottom left of the panel using a broad badger brush; strike the glaze across the direction of the grain pattern. Do not soften too uniformly. The process should be one of getting the right feel to the grain — you can develop certain areas by softening them out completely, whereas you can leave other sections quite sharp and graphic. Leave to dry for 24 hours.

4 Spatter small spots of dark brown or black oil-based glaze onto the panel to give texture. Do this by dragging your thumb or forefinger over the bristles of a stencilling brush that has been charged with glaze. This will release small drops of glaze onto the surface. Their disposition over the panel should be irregular so you should vary the technique as you work across the panel. While the glaze remains damp, soften and elongate the dots with a small badger brush. Work systematically, brushing along the direction of the underlying pattern.

5 A close-up detail reveals the elongated structure of the pores in the lighter-coloured areas of the panel. Allow to dry for 24 hours.

6 Seal with two coats of gloss or mid-sheen polyurethane varnish.

A S H

European ash is light gray to creamy white. It has a medium-coarse texture and straight, well-compacted grain. Differences in the tonal values of early and late seasonal growth produces decorative figures in plain and radial sawn timbers. American ash has similar figuring and grain structure, but varies in colour from gray-white to light tan.

Burr wood is common and may show pronounced or delicate markings. Tiny eyes appear as dark pin-heads of colour surrounded by swirling grain. The effect may be continuous through the cut or may appear as isolated clusters separated by plain wood. Fungus invasion may also cause pigment figures to appear as fine dark streaks.

Green ebony is the name applied to fallen ash that has been attacked by fungus; it is highly prized as a veneer.

Hungarian ash is pale yellow to light tan in colour, and exhibits a bold, twisted grain figure which produces a wavy, lustrous effect in plain sawn varieties. Rich and lively flame figures are also common.

Ash is something of a visual hybrid, with its rugged figure and creamy mild tones fitting in somewhere between the light oaks and pale pines. Its signature may be grained on furniture, panelling and architectural detail in any area within the home that requires a mild-mannered, uninsistent timber.

European ash

1 Mix a ground colour of bamboo eggshell finish tinted with a little white artist's oil colour to make a creamy yellow beige. Using a standard decorators' brush, apply this onto a previously prepared surface (see page 62) and allow to dry for 24 hours. In this example, the effect of two adjacent boards is created, so divide the panel into two by drawing a vertical ruled pencil line. (The figuring will be painted in before the background colour, so you do not need to mask out one half of the panel as in some other two-part simulations.) Oil the surface by using a clean lint-free cloth to wipe a thin layer of transparent oil glaze over the entire panel.

2 Using artist's oil colours prepare a transparent mid-brown glaze by blending raw Sienna, raw umber and a little black with white spirit. Take up the colour on a small flat fitch brush and paint in the grain pattern using the italic quality of the brush. Treat the two boards separately and construct patterns that indicate their individual nature. Soften the pattern with a small badger brush as you go along.

3 Soften the grain figure inwards towards the centre of the heart, using the small badger brush. This technique produces the characteristic tonal gradation in the figure, with the softened line hardening off to the dark, ragged, inner edge. Angle the badger brush in order to soften the compressed parallel lines of the grain. When you have achieved the desired effect allow the panel to dry for 24 hours.

4 Prepare a dark transparent glaze by blending raw umber, raw Sienna and black artist's oils with white spirit and a little transparent oil glaze. Using a 2.5 cm (1 in) decorators' brush, apply the glaze unevenly to the panel to create a series of toned areas that reinforce the composition you have already created and still provide the illusion of the two separate planks of wood.

5 Wipe out some areas of glaze to produce mottles, using the long, clean edge of a pencil eraser. These

mottles are visible because of light reflection and they usually occur where the tight, parallel compressions in the growth ring figure make sudden changes in direction. Again, remember to treat the two planks separately when locating the mottles.

6 Soften the mottles by striking them across their length with the small badger brush. At the same time soften any other strident tones that occur on the panel.

7 While the glaze remains workable, wipe out the inner edges of the heart grain with a cotton bud, using the dark edge of the underlying line as a guide. This light area is clearly visible in the natural wood and the effect of imitating this contrast of light and dark in paint is to strengthen the figure.

8 Lightly flog the surface of the panel to introduce the characteristic fine, compact grain structure. Work in vertical bands, moving systematically across the surface. Leave the panel for 24 hours to dry thoroughly.

9 Seal and protect the panel with two coats of mid-sheen polyurethane varnish. The finished composition is clearly defined by the contrasting figuring of the two boards. You can strengthen the junction between the two boards by outlining the join. To do this, use either black artists' oil paint on a fine artists' brush or a fine black marker pen. This optional process must precede the varnishing stage.

Note: Although the simulation demonstrates how the effect of two adjacent boards may be created, other combination designs can be achieved quite easily by first drawing the required pattern onto the prepared panel, and then following the basic procedures set out in the step-by-step instructions. Because there are no complicated and time-consuming masking-up procedures, the technique is particularly useful where intricate or repetitive designs are required.

VARIATION

In this variation, a featured panel of European ash has been enclosed within a mitred border. The finished result is shown in view **B** below. The design is first constructed within the border using an angled fitch. When dry, the borders are grained (as shown in view **A**). Patented graining tools are used to create a variety of heartwood figures in the wet glaze. The plainer sections are dragged diagonally with fine combs, to produce delicate interference patterns. Visually, these appear as mottled areas.

A

B

SATINWOOD

One of the classic cabinetmaking woods, satinwood is predominantly yellow in colour and is prized for its beautiful, clean finish. It is generally used as a contrasting backdrop to the more strident mahoganies, and is renowned for its ability to complement all manner of surfaces. West Indian satinwood was employed in England one hundred years before mahogany became the popular medium for the manufacture of furniture. It varies in colour from a pale yellow to a yellow-tan and is most prized for the clarity of its crotch and swirl figures.

East Indian satinwood comes from Sri Lanka and Southern India, and is light to dark golden-yellow in colour. It is fine, long and uniform in texture with indistinct pores and obscure rays. Although the heartwood is rather plain, in quarter sawn material the narrowly interlocked grain produces all the classic variations on stripe and broken stripe figures such as block mottle, raindrop, rope and fiddle back. The stripes are liberally scattered, with delicate cross mottles, and the wood shows a remarkable lustrous quality. With age, the wood mellows and darkens to resemble some of the paler mahoganies.

Satinwood is a quiet, unpretentious wood whose light, reflective surface finds many applications within the home. Its character is both sleek and sophisticated and warm and friendly. The various stripe figures may be presented as features or backdrops, and large areas may be grained without the effect becoming strident or overbearing. Crotch and swirl figures of varying sizes may be presented as either featured panels or focus designs on wall or furniture surfaces. In veneer form, the reflective qualities of the wood have been used to good effect in emphasizing the sculptural quality of curved or rounded details on furniture, walls and architectural features.

Stripe figure

1 Onto a previously prepared surface (see page 62), using a standard decorators' brush, apply a creamy yellow eggshell finish ground coat. Allow to dry for 24 hours. Prepare a dirty yellow glaze by mixing raw Sienna artist's oils with a little yellow ochre and Van Dyck brown. Blend this with white spirit and transparent oil glaze to achieve the required transparency and apply it to the panel with a 5 cm (2 in) decorators' brush. Brush-grain in a vertical direction and develop a series of slightly undulating tonal stripes of varying widths in the glaze. Keep changing the composition of the glaze as you work to give variations in the bands of tone.

2 While the glaze is still wet, immediately soften the stripes with a badger brush. Work in a predominantly horizontal direction, moving systematically across the panel. Use the brush with a light flicking motion, applying sufficient pressure to make the bristles flex.

3 Prepare a deep red-brown transparent glaze by blending burnt Sienna artist's oil colour with white spirit and a little transparent oil glaze. Take up the glaze on a medium fitch brush and strengthen the existing pattern by introducing thin irregular stripes onto the panel with the edge of the brush. Position them randomly and sparsely across the panel.

4 Lighten and soften the entire surface of the panel once more, using the badger brush. At this stage you can introduce further variations in both the width and definition of the stripes.

5 Although satinwood is fine-grained and evenly textured, you can introduce fine pores onto the panel by lightly flogging the glaze with the edge of a fan fitch brush. Work systematically down the panel, concentrating on one stripe at a time and paying particular attention to the darker stripes. Do not overdo this or you will lose the clarity of the stripes. Allow to dry for 24 hours.

6 Seal and protect with two coats of mid-sheen polyurethane varnish.

Heart figure

1 Using a standard decorators' brush, apply a creamy yellow eggshell-finish ground coat onto a previously prepared panel (see page 62). Allow to dry for 24 hours. Then, using gouache colours, prepare a pale yellow transparent glaze by blending raw Sienna with a little yellow ochre. Apply the glaze quickly and evenly with a wide brush. Make broad horizontal sweeps across the surface of the panel.

2 Immediately stipple the entire surface of the panel with a badger softening brush to level out the glaze and disguise any brush marks. By concentrating the stippling in selected areas you can introduce slight variations in the distribution of the coloured glaze across the surface.

3 Soften the surface, working in all directions with a large badger brush. The slight unevenness in the underlying glaze will produce pleasing tonal variations in the ground.

4 Wipe out mottles in the wet glaze, using the sharp edge of a pencil eraser. Push the excess colour into the lower sections of the mottles. Angle and drag the eraser to produce the required variations in the mottle shapes. The exact positions of the mottles depends on your chosen composition. In this case the panel has been designed to display the characteristic mottled highlights that occur as the heart-wood figure begins to change into a more regular striped appearance. (For the stripe figure simulation see page 95.) The mottled pattern is symmetrical about the long central axis of the panel and looks rather like a cloudy sky.

5 Soften the panel with a large badger brush by striking the surface of the wet glaze at right-angles to the directional flow of the mottles – in other words, work in towards the centre of the base of the panel. Farther out, the concentrations of displaced colour that lie in the bottoms of the wiped-out sections should be feathered out in fine tonal gradations. This will produce the

7

8

9

characteristic 'lights and darks' of the mottle figure. When you are happy with this effect, allow the panel to dry out thoroughly.

6 Using artist's oils prepare a deeper yellowy-brown glaze by mixing raw Sienna with small quantities of yellow ochre and Van Dyck brown. Reduce with white spirit and a little transparent oil glaze to give the necessary transparency, and use a 2.5 cm (1in) decorators' brush to apply the glaze to the entire panel in a loose irregular fashion.

7 With the decorators' brush indicate the rough form of the heart figure by brush-graining on some of the same deeper yellowy-brown glaze used in Step 6. Apply it in a series of arcing loops.

8 Prepare a red-brown transparent glaze by blending burnt Sienna artist's oil colour with white spirit and a little transparent oil glaze. Take up the colour on a small flat brush and apply it to selected areas of the heart grain and stripes to strengthen the pattern.

9 Now drag a modified wallpaper-hanging brush (see page 80) through the glaze in a series of long continuous sweeps. The colour will separate into the finer and more distinctive parallel banding of the heart figure.

10 Use a large badger brush to soften and merge the coloured pattern. The softening technique is achieved by striking out from the centre of the heart towards the top and sides of the panel. Finally soften the stripes across their long axis.

11 Introduce fine pores onto the panel by gently flogging the surface with the edge of a small badger brush. The disposition of pores should be fairly regular but they may be concentrated in the darker areas of the figure. Although in some instances the pore structure may relate to the flow of the grain, on the whole it should describe a steady vertical course up the length of the panel. Allow the panel to dry for 24 hours.

12 Seal and protect the finished panel with two coats of mid-sheen or matt polyurethane varnish.

10

11

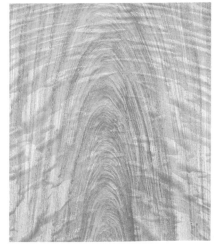

12

Block mottle figure

1 Onto a previously prepared surface (see page 62), using a standard decorators' brush, apply a ground coat of creamy yellow eggshell paint. Allow to dry for 24 hours. Using artist's oils, prepare a mid-yellow glaze by mixing raw Sienna with small quantities of yellow ochre and Van Dyck brown. Blend with white spirit and a little transparent oil glaze to achieve the required transparency. Take up the glaze on a 5 cm (2 in) decorators' brush and apply the colour to the entire panel in a loose, irregular fashion.

2 Using a long-bristled overgrainer, break up the wet glaze into a series of short, vertical stripes of varying tones and lengths. It is best to use the brush with a pushing or 'pouncing' motion. The outer edges of the stripes will be formed into a series of wispy ragged serrations. If you use your brush alternatively in your right and left hand, you can use it to produce this effect on both sides of the stripe.

3 Lightly soften the entire panel in a horizontal direction, using a small badger brush. Work from left to right and then from right to left, to merge the lines into a series of tonal zigzags. The effect should not be uniform – vary the amount of softening over the panel to produce a range of different tones. When you are happy with the effect, leave the panel for 24 hours to dry out thoroughly.

4 With the 5 cm (2 in) decorators' brush, brush-grain the entire surface of the panel into a series of fine stripes, using the glaze prepared in step 1. Slightly varying the composition of the glaze throughout the process will ensure the necessary tonal range.

5 While the glaze is still wet, lightly flog the surface of the panel with the edge of a small badger brush. The pored effect is not uniform, so vary the amount of flogging you do as you work over the surface. Work down the panel in vertical bands. Allow to dry for 24 hours.

6 Seal with two coats of matt or mid-sheen polyurethane varnish.

Rope figure

1 Using a standard decorators' brush, apply a ground coat of creamy yellow eggshell-finish paint onto a previously prepared surface (see page 62). Allow to dry for 24 hours. Using artist's oil colours, prepare a mid-yellow glaze by mixing raw Sienna with small quantities of yellow ochre and Van Dyck brown. Blend with white spirit and a little transparent oil glaze to achieve the required transparency. Take up the colour on a 5 cm (2 in) decorators' brush and brush-grain the panel into a series of roughly vertical tonal stripes. Make sure there is plenty of tonal variation.

2 Using a long-bristled overgrainer, create a series of mottles by concentrating the wet glaze into short, diagonal lines. Use a pushing or pouncing action and work in one direction only. The object is to leave a build-up of glaze along the bottom edge of each mottle. Although the stacked mottles may vary in width, they should be in recognizable vertical columns. Work in bands from top to bottom of the panel.

3 Gently soften the mottles using a small badger brush in a vertical direction. Work systematically over the panel, and vary the technique to ensure a variety of different tones.

4 While the glaze remains alive, lightly drag an overgrainer through the columns in a roughly vertical direction. The technique should be continuous and you should swerve and hesitate as you drag the brush in an attempt to follow the lights and darks of the stacked mottles.

5 The dragging technique used in Step 4 tends to have a rather severe effect. You should therefore lightly flog the surface in vertical bands using the edge of the small badger brush. This should break up the continuity of line. Allow to dry for 24 hours.

6 Seal with two coats of matt or mid-sheen polyurethane varnish.

Note: This simulation has been described using artist's oil colours. Alternatively you can carry it out in water colours, in which case drying times are far quicker.

Raindrop figure

1 Onto a previously prepared surface (see page 62), apply a ground coat of creamy yellow eggshell-finish paint with a standard decorators' brush. Allow to dry for 24 hours. Prepare a mid-yellow glaze by mixing raw Sienna with small quantities of yellow ochre and Van Dyck brown artist's oils. Blend with white spirit and a little transparent oil glaze to give the required transparency. With a 2.5 cm (1 in) decorators' brush, brush-grain the panel to create a series of slightly undulating parallel bands. You can introduce the tonal variations required for this type of grain by altering the composition of the glaze throughout the process.

2 Lightly stipple the entire surface of the panel with a jamb duster brush to soften and merge the bands. Vary the technique so that some of the areas of banding are left distinct.

3 With an overgrainer, use a pushing action to create a series of variable-sized diagonal mottles. The pattern should be neither regular nor strictly diagonal. Work systematically over the surface and concentrate the raggy areas of excess glaze into the bottoms of the mottles.

4 While the paint remains mobile, pull a modified wallpaper-hanging brush gently through the glaze. Hold the brush at the same angle as the mottles and pull it down and slightly across the panel in a continuous hesitating and swerving movement. Repeat the process over the rest of the panel. Some variation in the dragging angle and the attitude of the brush is necessary to prevent the pattern from looking mechanical or contrived.

5 Soften the pattern both along and across the diagonals of the mottles with a small badger brush. Vary the technique to ensure that parts of the panel remain sharp while others are well softened. Allow the panel to dry for 24 hours.

6 Seal with two coats of matt or mid-sheen polyurethane varnish. Although the finished raindrop figure shows a strong diagonal accent, a slight wave or swirl in the overall pattern gives the best compositon.

ROSEWOOD

Among the many varieties, the Brazilian and East Indian rosewoods are the most decorative. East Indian rosewoods are found in East India and Sri Lanka. They come from the blackwood tree, and are medium to dark purple-brown, marked with dense black streaks. The wood is even-textured, and although possessing an interlocked grain structure, gives only indistinct stripe figures. Unusually, the best figures are produced by cutting at right angles to the growth rings. The wood takes a smooth, lustrous finish and is renowned for the manner in which it complements fabric.

Brazilian rosewood is yellowish-tan in colour, but this may vary from orange, through deep red tones to dark purples. It shows dark pigment figures, especially when irregular logs are plain sawn. It is medium-coarse in texture, shows scattered vessel lines and, in some instances, exhibits wavy-grain characteristics.

Rosewoods are most famed for their intricate and complex figuring patterns. While there is a definite directional flow along the length of the tree, the heartwood may be seen to be composed of a series of irregular, disjointed sections. Each shows its own distinctive figuring characteristics and each is separated from its neighbours by dark sketchy lines or smudged areas of pigmentation.

The dark formality of the wood is beautifully counterbalanced by the abstraction of the intricate figuring. The wood has particular application to fine, delicate pieces where the eye is drawn onto the surface and then invited to browse among the exotic patterns. For contemporary themes, the bizarre figuring provides an endless source of inspiration in the creation of fantasies.

Rosewood figure

1 Mix a ground colour of eggshell-finish in nectarine strengthened with a small amount of deep red oil colour and use a standard decorators' brush to apply this to a previously prepared surface (see page 62). Allow to dry for 24 hours. Using gouache colours, prepare a deep red glaze by mixing Van Dyck brown and Indian red. Thin this glaze to transparency with water. Brush-grain the panel, using a 2.5 cm (1 in) brush, to produce a series of broad bands. The size and direction of these bands should be governed by the position of the main features in the finished panel – in this case the heavily figured central heartwood and the plainer surrounding areas of sapwood (see Step 12).

2 With the same brush, push or 'pounce' the glaze in the areas of the sapwood to produce a series of loose tonal irregularities and mottled features. Their accent should lie mainly across the long axis of the brush-grained pattern and their position should relate to distinctive features in the grain of the sapwood.

3 Using a large badger brush, lightly drag the glaze in a series of vertical bands, working systematically across the panel. The mottles and tonal variations will be softened and the long grain axis of the wood will be re-established. When you have done this, allow the panel to dry out completely.

4 Prepare a mid- to deep-brown oil-based glaze by mixing Van Dyck brown with small quantities of black artist's oils, then thin to transparency with white spirit and a little transparent oil glaze. Use the 2.5 cm (1 in) decorators' brush to brush-grain the central area of the panel that is to represent the heartwood.

5 Using a fan badger brush, make a series of swirling patterns in the glaze. These will provide the foundation for the detail that is to be superimposed in the next step. If you are unsure of the unusual grain formation of rosewood, refer to a sample, sketch or photograph of the natural wood when carrying out this

7

10

and the subsequent stages.

The fan badger brush is capable of producing a vast range of parallel marks if you angle it, rotate it, and vary the pressure on the head of the brush as you work.

6 Soften and texture and glaze using the edge of a small badger brush.

7 Deepen some of the glaze you prepared in Step 4 by adding further amounts of black. Take up the glaze on a small flat fitch brush and, using the italic qualities of the brush, paint in the heart-grain figure. Use the underlying patterns as a compositional guide.

8 With the glaze prepared in Step 4, brush-grain the areas that surround the central heartwood. Drag the 2.5 cm (1 in) decorators' brush in a predominantly vertical direction down the panel, following the wavy contours of the heartwood.

9 Drag the glaze into a series of undulating parallels with a fan badger brush. Use the mottles and toned areas created in Steps 2 and 3 to indicate where the kinks and swerves in the grain direction should occur.

10 Take up the glaze prepared in Step 7 on the edge of the flat fitch brush and strengthen the definitions between the distorted enclosed formations of the heartwood figure.

11 Using the edge of a fan badger brush, lightly flog the entire panel to re-establish the fine, grainy texture of the wood. Allow the panel to dry for 24 hours.

12 Seal and protect with two coats of gloss or mid-sheen polyurethane varnish. The finished panel shows the disjointed complex figuring of the rosewood heart, which is unique and unmistakable. The more plainly figured, surrounding sapwood gives a contrasting backdrop to highlight the more decorative centre.

Note: This simulation has been described as a two-stage process, beginning in water colours and finishing in artist's oils. You can carry out the whole simulation in oils, although the drying time allowed for the panel at the end of Step 3 will then be 24 hours.

8

11

9

12

OAK

Among the various species of oak native to Europe, the English oak is the most famous, by virtue of its association with the early history, literature and arts of the country. And because of its performance as a durable, stable and decorative wood, its name has become synonymous with strength and reliability. Indeed, so popular was English oak — as opposed to imported woods from Spain, Turkey and Austria, which were of an inferior quality and figuring — preservation orders were being placed on the rapidly dwindling stocks as early as the sixteenth century.

English oak varies in colour from a light yellow-tan through to deeper and warmer shades of brown. It is coarse-textured and straight-grained, displaying distinctive growth rings in plain sawed material. Rays are of two types; in plain sawed timbers they appear as dark lines, often exceeding 1 in (2.5 cm) in length; whereas in quarter sawn timbers they appear as lustrous, broad silver flecks, which form into the flowing patterns of the characteristic, and highly decorative, 'flake figure'.

European and Russian oak is similarly coloured, and displays lively heart and flame figuring, large silver rays, and is virtually free of knots.

All oak timbers weather naturally to a pale gray colour; the wood only attaining its familiar range of yellow and brown colouring through the successive application of oils and varnishes. Indeed, the wide variations of tone and colour, coupled with the natural warmth of the material, allows the grainer many opportunities for incorporating an oak simulation within the home.

Quartered oak (flake figure)

1 Mix a ground coat of a bamboo colour in an eggshell finish, blended with a little white artist's oil colour to a creamy beige. Use a standard decorators' brush to apply this onto a previously prepared surface (see page 62) and allow to dry for 24 hours. Using artist's oil colours, prepare an oil-based glaze by mixing raw Sienna, raw umber and yellow ochre with a little black. Thin this with white spirit and a little transparent oil glaze. With a 2.5 cm (1 in) decorators' brush take up this glaze and brush-grain the panel in a series of fine, slightly undulating, stripes. Vary the tone of the stripes by adjusting the glaze mix.

2 Using the edge of a small badger brush, gently flog the surface of the glaze to soften the stripes and give texture. Work down and across the panel, flogging the bands diagonally.

3 With the short sharp edge of a pencil eraser, wipe out the flakes, displacing excess glaze onto the lower edge of each individual section. When you are satisfied that it is as natural-looking as possible, allow the panel to dry out thoroughly for 24 hours.

4 Prepare a dark transparent glaze by blending burnt umber and black artist's oils with white spirit. Using the 2.5 cm (1 in) decorators' brush, apply the glaze unevenly to the panel to create a series of light- and dark-toned areas. Softening the flake figure itself would disguise the sharpness of the feature. The object of the glaze is to provide an overall softening of the pattern by harmonizing the tonal contrasts.

5 While the glaze is still tacky, flog the surface using the edge of a long-bristled flogger. This technique softens the tonal gradation of the applied colour and also textures the surface. As the glaze hardens off, further flogging will result in the displaced glaze being taken up on the bristles of the flogger and redeposited as a series of fine broken lines or elongated pores. Allow the panel to dry for 24 hours.

6 Seal and protect the finished panel with two coats of matt or mid-sheen polyurethane varnish.

Dark oak figure

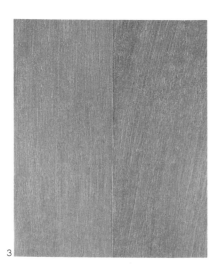

1 Using a standard decorators' brush, apply a ground coat of eggshell-finish in a bamboo shade onto a previously prepared surface (see page 62). Allow to dry for 24 hours. Draw a vertical ruled pencil guideline down the middle of the panel. Make the oil-based ground ready to take water colours by rubbing a small quantity of fuller's earth powder over the surface of the panel with a damp cloth. Using gouache colours, prepare a deep brown-yellow glaze by mixing burnt umber, raw umber and black. Lay a straight edge along the guideline that halves the panel and use a 2.5 cm (1 in) decorators' brush to apply the glaze loosely, brush-graining each side of the panel separately.

2 Again using the straight edge to separate the two halves of the panel, lightly flog each side with the edge of a small badger brush, striking the surface at a rough diagonal to the long axis of the grain. Leave the panel to dry out.

3 The different brush-graining and flogging procedures that have been carried out independently on the panel in Steps 1 and 2 have created a composition which simulates two separate boards, giving the effect of a simple quartering or panelling technique.

4 Prepare a dark brown glaze by blending burnt umber and black artist's oil colours with white spirit and transparent oil glaze. Use a 2.5 cm (1 in) decorators' brush to apply the mixture evenly to the entire panel. As well as toning the panel, the glaze will permit the superimposed figuring details to be softened in the required way.

5 Deepen the glaze prepared in step 4 by adding more black, and reduce it further with white spirit. Take up the glaze on a flat fitch brush and figure each side of the panel in a contrasting pattern. Use the italic quality of the fitch brush to create lines of varying thicknesses in both the heart and the surrounding wood. Soften the details with a small badger brush as you proceed.

6 The technique of softening involves striking the figured line in

towards the centre of the heart. The white spirit in the figuring glaze eats into the underlying colour and reveals the lighter-coloured ground. Softening produces a gradual tonal progression across the line which eventually culminates in a thin, dark, hard edge forming on the inner side of the figure (see Step 11).

7 Lightly flog the figuring with the edge of a small badger brush to break up the continuity of the line. Work systematically down the panel, striking the figure diagonally.

8 Use an overgrainer to reinforce the coarse long-grained textural qualities of the timber. Drag the brush in a continuous vertical motion down the entire length of the panel. Work systematically across the surface but vary the technique slightly for each board. The dragged effect is clearly evident in the heavily serrated arcing elipses of the flame-like figure. When you have achieved a satisfactory result, allow the panel to dry for 24 hours.

9 Prepare a thin black water-based glaze from black gouache colour. Blend the paint with a pinch of fuller's earth powder and, using a 2.5 cm (1 in) decorators' brush, apply the colour quickly to the entire surface. The object of this glaze is to tone and harmonize the various colours on the panel. Try to get an uneven distribution of colour, remembering to preserve the separate identity of the two boards.

10 Gently flog the lighter areas of the panel with a fan fitch brush. The sparse splayed bristles produce a variety of elongated 'pores'. Work in vertical bands to re-establish the long-grained texture of the wood and remember once more to slightly vary the technique between the two boards. Allow to dry for 24 hours.

11 This detail shows how the heavily fractured lines and carelessly scattered, dark, elongated pores produce a coarse, long-grained, textural quality in the surface. The serrated edges of the softened figure are clearly evident in the central area of the crown.

12 Protect with two coats of matt or mid-sheen polyurethane varnish.

Light oak figure

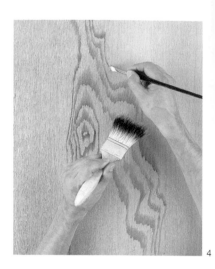

1 Use a standard decorators' brush to apply a ground coat of eggshell-finish in bamboo onto a previously prepared surface (see page 62). Allow to dry for 24 hours. Prepare the oil-based eggshell ground to accept water colours by rubbing a small quantity of fuller's earth powder onto the surface with a damp cloth. Using gouache colours, make up a mid-brown glaze by blending raw umber with a little black. Thin this glaze to transparency with water. With a 2.5 cm (1 in) decorators' brush, brush-grain the surface to create a series of vertical tonal stripes.

3 Grain and texture the surface by flogging the panel with the broad, flat face of a long-handled flogger. Work in a series of vertical bands across the surface of the panel. Vary the technique to give the surface some tonal variation. Allow the panel to dry.

3 Oil in the ground by wiping a thin layer of transparent oil glaze over the entire surface of the panel with a clean cloth. This provides a lubricating base on which superimposed glazes may more easily be worked and softened.

4 Prepare an oil-based mid-toned brown glaze by blending raw umber artist's oil colour with white spirit and a little transparent oil glaze. Take up the glaze on a flat fitch brush and using the italic potential of the brush begin by painting in the heart grain. As you go along, use a small badger brush to soften inwards towards the enclosed heart. The technique produces a soft tonal gradation of line which hardens off to a fine sharp edge on the inner side of the figure.

5 Allow the panel to harden off a little before fracturing the completed figure with the edge of the long-handled flogger. Work systematically over the panel striking the surface at a diagonal to the length of the panel. Allow the panel to dry completely.

6 Introduce the distinctive long pores onto the panel using a check roller. Prepare a black oil-based glaze by blending black artist's oil

colour with white spirit. Take up the glaze on a 2.5 cm (1 in) decorators' brush and apply it to the wheels of the roller as you push it over the surface. Work mainly in a vertical direction and take care not to overstate the effect or introduce too regular a print onto the surface. Allow the panel to dry thoroughly for 24 hours.

7 Prepare a dark brown-black oil-based glaze by blending black and burnt umber artist's oils with white spirit. Using a 2.5 cm (1 in) decorators' brush, apply the glaze to the entire panel. Try to distribute the glaze unevenly, to produce the required tonal variation on the surface.

8 Flog the entire panel with the edge of the long-handled flogger. Work in a series of vertical bands, going over the surface systematically. The tonal areas will be softened and merged, and the whole panel will take on the characteristic grainy appearance of oak. Vary the technique, applying lighter and heavier flogging to different parts of the panel, to ensure a natural look. Allow to dry for 24 hours.

9 Seal and protect with two coats of matt or mid-sheen polyurethane varnish. The finished panel shows the lively flame-like figure struggling to escape the tight compressions of the surrounding sapwood. To avoid a contrived appearance, the heart is best positioned away from the centre. Displays like this are highly decorative and may be treated as featured panels within a contrasting border of plainer oak.

Note: Even when experienced grainers are completely familiar with the various patterns found in the natural material, when preparing a panel the majority still prefer to work from a sketch or sample. This is particularly important when graining a feature panel in which figuring detail produces the main compositional elements. In some cases the pattern may be drawn onto the panel in pencil before painting.

VARIATION

A Main variation: This panel features a matt gold ground overgrained in a blend of raw umber and burnt Sienna. Although oil colours were used, water-based colours are equally effective. The technique is fast, and involves graining the surface with a patented graining tool and a selection of combs (see p. 67).

B Detail of A: This detail shows the graphic print of the graining tool used in creating the heart figure. Mottles are made by combing diagonally across the grain's long axis.

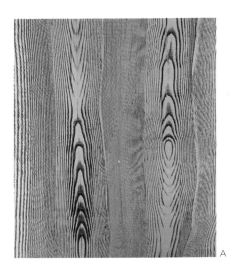

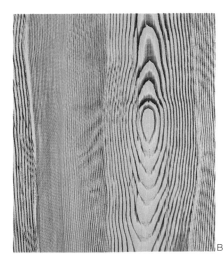

Burr oak figure

1 Onto a previously prepared surface (see page 62), use a standard decorators' brush to apply a eggshell-finish bamboo-coloured ground coat. Allow to dry for 24 hours. Prepare an oil-based glaze by blending burnt umber artist's oil paint with white spirit and a little transparent oil glaze. Using a 2.5 cm (1 in) decorators' brush, apply the colour to the panel in a series of loose irregular broken lines.

2 Using a modified wallpaper-hanging brush (see page 80) with sparse irregular bristles, drag the glaze into a series of continuous, swirling patterns and complex, enclosed shapes. The disjointed nature of the initial rubbed-in colour makes many unusual and surprising variations possible in the dragged pattern. You must recognize and develop these at this stage.

3 Soften the panel by striking the surface with a large badger brush. As you work across the whole surface, vary the technique to ensure a variety of effects.

4 Using a small fitch brush, produce a series of variable-sized circles in the wet glaze. The technique is achieved by spinning or rotating the shaft of the brush in the hand while the bristles are in contact with the surface of the panel. Angling the brush and varying the pressure on the bristles permits a wide range of different-sized strokes and effects. Organise the circles into groups, which should be surrounded by swirling grain formations.

5 Prepare a slightly deeper, more opaque version of the original glaze. Take up the colour on a small fitch brush and dot it into the centres of the dragged circles by rotating the shaft of the brush in the same way that you did in Step 4. Adjust the technique to vary the visual impact of the circles.

6 Soften the circles in all directions with a small badger brush. Develop any promising details that emerge.

7 Deepen the glaze used in Step 5 by adding more black. Take up the glaze on an artist's pencil brush and strengthen the centres of the circles by dotting on a little glaze. Once

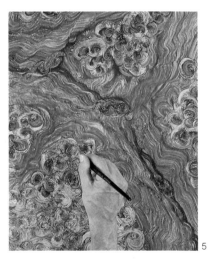

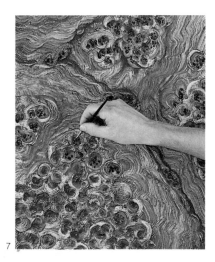

more, try to vary the technique – heighten some of the more clearly defined circles, but leave others quite soft and understated. Such variety will enhance the depth of the finished panel and make it more realistic.

8 Complete the process by flogging the channels with the edge of a small badger brush. Vary the direction of the strokes, while using the technique to produce a result that complements the natural flowing quality of the grain. Allow to dry for 24 hours.

9 Seal and protect with two coats of matt or mid-sheen polyurethane varnish. The finished panel shows the essential qualities that are common to all burr, root, pollard and stump cuts. Closely grouped, complex eye details are surrounded and enclosed by swirling grain formations. You can change the scale and size of these elements to suit the area in which the finished panel will be displayed.

VARIATIONS

A To the basic burr background, constructed in Steps 1–4 shown on the opposite page, the grainer has added elongated pores in a similar manner to bird's eye maple graining (see the finished bird's eye result shown in Step 8, on page 121), using black artists' oils on a fine brush.

B The burr pattern can be very effective for graining panels or marquetry when executed as a double book-end pattern (see page 54).

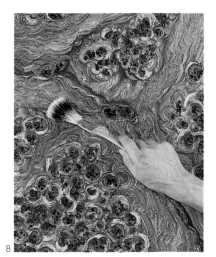

7

8

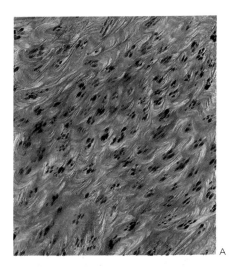

A

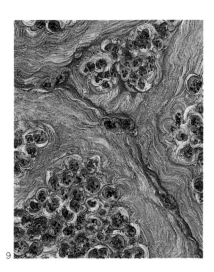

9

Note: This simulation is carried out in artist's oil colours. It is possible to use water colours, but the graining process is quite lengthy and there is the possibility that water colours will dry out too quickly.

Although burr cuts are prized for their exotic and irregular figures, when grained panels are prepared as features, the composition is usually formalised into a balanced book-end or double book-end focus pattern (see Variation B, right). When preparing such panels, the major elements in the composition should be clearly drafted out onto the prepared surface before applying colour.

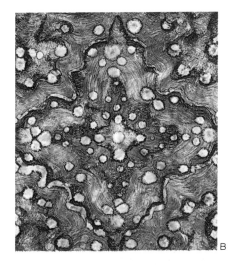

B

WALNUT

The principal cabinetmaking wood in Italy in the fifteenth century, walnut's popularity spread to Northern Europe, where the various Persian, Turkish, Circassian, Italian, Spanish, French, Rheinish and English varieties became recognized for their individual qualities. Before supplies became limited it was regarded as a general-purpose utilitarian timber that was suitable for carving and turning. However, it was subject to worm attack and after 1740 it fell out of fashion and was largely superseded by the more resilient mahoganies. Interest was revived in the early and mid part of the nineteenth century, when it was used in the form of veneer to decorate some of the most elegant furniture.

Walnut has a medium-coarse but uniform texture, and is straight-grained. It exhibits light to very dark gray-brown tones, and when quarter sawn is capable of producing stripe, roe, wavy and mottle figures. In veneer form, walnut burr is much prized for its great depth and subtlety of pattern.

Although the lighter beige and fawn varieties may be described as warm, on the whole walnuts tend to be rather formal and sophisticated. Their subdued tones are certainly more at home in the refined atmosphere of some generously proportioned drawing room than the enforced intimacy of a small flat. Although lively burr cuts and 'butty' figures provide splendid focus patterns or featured panels on wall and furniture surfaces, over large areas the conservative character of the wood is generally reinforced by displaying the material in a regimented striped format. The formality of the wood does however provide a pleasing contrast to livelier, lighter-coloured timbers such as satinwood or yellow maple, and it is in the form of parquetry or marquetry combinations with these woods that walnut finds its most appropriate expression in the modern interior.

Burr figure

1 There is much variation in the colour of the various walnuts and you can use cinnamon, brownstone or nectarine eggshell-finish paint as a ground. Using a standard decorators' brush, apply your chosen ground coat onto a previously prepared surface (see page 62). Allow to dry for 24 hours. As a precaution, rub a little fuller's earth powder onto the panel with a damp cloth. This will prevent the water-based glaze from 'cissing' or retreating from the point of application of the brush. Using gouache colours, prepare a mid-brown glaze by blending Van Dyck brown with a little black. Thin with water and apply quickly and evenly to the panel using a broad flat brush.
2 Using a modified wallpaper-hanging brush, drag a series of random overlapping swirls into the wet glaze and allow the panel to dry.
3 Prepare a deeper glaze in artist's oils by bending Van Dyck brown and black with white spirit and a little transparent oil glaze. Using a 5 cm (2 in) flat brush loosely apply the colour in a series of broad, continuous bands.
4 Using the same brush slightly moistened with white spirit, push the bands of wet glaze into a series of distorted hollows. The presence of small amounts of white spirit causes the dark pigment to ciss. The underlying pattern is revealed and you can use the swirls to suggest where to place further hollows or shapes.
5 Make a thin transparent glaze by blending neat black artist's oil colour with copious amounts of white spirit. Take up the colour on a flat fitch brush and introduce it onto the panel into the middle of the hollows. As you dot the glaze onto the panel a more pronounced cissing will cause the displaced colour to pool into darker, stronger circles. The amount of solvent on the brush and the length of time it remains in contact with the surface will affect the size and form of the circles. Allow to dry for 24 hours.
6 Seal and protect with two coats of mid-sheen polyurethane varnish.

Stripe figure

1 Using a standard decorators' brush, apply a ground coat of cinnamon, brownstone or nectarine eggshell-finish paint onto a previously prepared surface (see page 62). Allow to dry for 24 hours. Using gouache colours, prepare a mid-brown glaze by blending Van Dyck brown with a little black and burnt Sienna. Rub a small quantity of fuller's earth powder onto the panel with a damp cloth to prevent the water-based glaze from cissing. Thin the glaze to the required transparency with water, and brush-grain the panel, using a 2.5 cm (1 in) decorators' brush, to produce a series of tonally varied, vertical stripes.

2 Stipple the entire surface of the panel with a jamb-duster brush to soften the striped effect. Vary the technique so that some areas are well stippled and softened while other areas still retain clear evidence of the stripes.

3 Wipe out the mottles using the long sharp edge of a pencil eraser. The mottles should be thin and scattered casually over the panel. They should lay mainly across the direction of the underlying stripe. As you work, displace the excess glaze into the lower portions of the mottles.

4 Soften the panel with a large badger brush. Although you can soften in all directions, concentrate on a downward accent, so that you produce the necessary tonal gradations in the bases of the mottles. Then allow the panel to dry.

5 Using artist's oil colours, prepare a deeper tonal glaze by mixing Van Dyck brown, burnt Sienna and a little black. Thin to transparency with white spirit, take up on a 2.5 cm (1 in) decorators' brush, and brush-grain the panel into a series of tonal stripes. Note that the angle of tonal stripes is slightly different from the angle of those produced in Step 1.

6 Soften the panel with a large badger brush, working across the direction of the stripes. Be systematic, but adjust the technique as you move over the surface to ensure a variety of softened strokes.

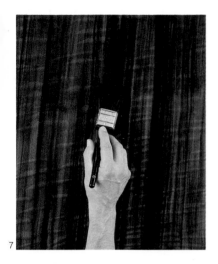

7

8

9

Allow the panel to dry for 24 hours.
7 Continuing in water colours, prepare a deep brown glaze by blending Van Dyck brown and black gouache colours. Thin to transparency with water, but before applying the glaze to the surface blend a pinch of fuller's earth powder into the mixture to prevent cissing. Using a 2.5 cm (1 in) decorators' brush, apply the glaze evenly over the entire panel.
8 While the glaze is still wet, texture the surface by flogging the panel with the edge of a flogging brush. Move systematically across the panel, working in series of vertical lines from the top to the bottom. Vary the effect by adjusting the technique as you work up and down the surface. Allow the panel to dry.
9 Seal the finished panel with two coats of mid-sheen polyurethane varnish. This example shows how, as with all striped woodgrains, it is desirable to arrange the pattern at a slight angle to the long edge of the panel.

Note: A walnut stripe is often used as a surrounding border to enclose more complex featured patterns. In this instance, the simulation of the stripe can be considerably simplified. The tonal brown glazes produced in steps 1 and 5 can be brush-grained (see *Dragging*, page 75) onto the panel in one operation. Softening may then be carried out with the large badger brush to create the required tonal gradations in the stripe figure. The whole panel can then be lightly flogged to texture the surface. If gouache (water-based) colour is used, the whole panel may be completed — and dry — in a matter of minutes. The striped figure shown in Variation B was created in this manner.

VARIATIONS

A Citrus yellow ground colour has been used in this fantasy striped variation. A pale red glaze is brush-grained onto the panel, lightly dragged and allowed to dry. The more opaque red overgraining colour is brush-grained into a series of tonal stripes, and then softened.

B Here the figuring is displayed on the diagonal, in opposite directions, to produce a striking sense of movement. A medium-brown glaze was brush-grained onto the panel, then softened to produce a quick but realistic result.

A

B

'Butty' figure

1 Prepare a ground coat of cinnamon, brownstone or nectarine eggshell-finish paint and brush it onto a previously prepared surface (see page 62), using a standard decorators' brush. Allow to dry for 24 hours. Using gouache colours, prepare a mid-brown glaze by mixing Van Dyck brown and burnt Sienna with a little black. Rub a small quantity of fuller's earth powder onto the oil-based ground on the panel with a damp cloth. This will prevent the applied glaze from cissing. Thin the glaze to the required transparency with water and, using a 2.5 cm (1 in) decorators' brush, brush-grain the colour onto the panel to produce broad undulating swirls of tone.

2 Using a modified wallpaper-hanging brush, strengthen the pattern by dragging the wet glaze into a series of continuous dipping and curving waves. Leave some areas of the panel relatively free of pattern, to contrast with the areas that contain more elaborate figuring.

3 Mottle the plainer striped areas on the left-hand side of the panel using the long, thin edge of a pencil eraser. Position the mottles so that they lie across the direction of the grain and displace the excess glaze down onto the lower edges of the mottles.

4 Soften the entire panel with a large badger brush. Begin by softening the mottles in a downward direction; then work upwards and outwards from the base of the panel, softening across the direction of the wavy pattern. Vary the technique so that some areas are totally merged while others remain quite sharp. Then allow the panel to dry.

5 Prepare a slightly deeper brown oil-based glaze by blending Van Dyck brown with burnt Sienna and a little black artist's oil colour. Thin to transparency with white spirit and a little transparent oil glaze and, using a 2.5 cm (1 in) decorators' brush, brush-grain the surface of the panel into a series of parallels that roughly echo the underlying pattern.

6 Drag the glaze once more, using a

modified wallpaper-hanging brush. Because the bristles are very sparse, splayed and uneven, the result is a variety of fine sketchy strokes. Use the underlying pattern as a rough guide only, and introduce further overlapping waves and swirls into the glaze.

7 After further softening, the gradual build-up of superimposed patterns begins to produce great depth in the panel. Vary the technique and develop any areas that show particular promise.

8 Now use a flat fitch brush to introduce stronger bands of darker colour into the wet glaze. Exploit the italic quality of the brush to help you vary the width of the line and use the underlying pattern to indicate the course of these stronger bands of colour. They should be roughly parallel but unevenly spaced to prevent the pattern appearing mechanical or contrived.

9 Soften the effect of the dark lines with a large badger brush, working up and out from the base of the panel. Again, vary the technique to create variety in the softened print. Leave the panel for 24 hours to dry.

10 Make a deep brown-black water-based glaze by mixing gouache colours in Van Dyck brown and black. Thin the glaze to the required transparency with water and blend it with a pinch of fuller's earth to prevent the colour cissing on the surface. Using a 2.5 cm (1 in) decorators' brush, brush-grain the panel in a direction that roughly echoes the underlying pattern.

11 Flog the panel with the edge of a long-handled flogger brush to texture and tone the applied glaze. Once again, work with the direction of the pattern. Allow the panel to dry for 24 hours.

12 Seal and protect with two coats of mid-sheen polyurethane varnish. The soft overlapping bands of the finished panel produce the characteristic subtlety and depth of the classic walnut grain. Although in this example the more elaborate wavy figure has been contrasted with a simple stripe, you can try other pattern combinations.

Quartering

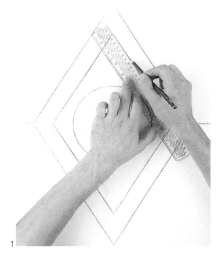

This sequence shows how you can combine a variety of simulated woods to produce a simple geometric pattern – an effect rather like painted marquetry. In this instance the woodgrains were painted on stiff plastic film. Individual sections were cut out, primed with their respective base colours, grained and then glued onto a solid surface. The woodgrains featured are walnut, oak burr, bird's-eye maple and mahogany swirl. Another way to achieve the same effect would be to work in a continuous sequence on a solid board. This would entail reversing the sequence of work, so that you would start with the central disc of mahogany and work outwards to finish with the border of walnut. You would also have to allow each section to dry thoroughly before going on to the next piece.

1 Draw up the pattern on a sheet of plastic film, using a pencil, ruler and compasses. Use a pair of scissors to cut out the different sections.
2 Paint the sections in their respective eggshell-finish base colours. These are cinnamon for walnut, bamboo for oak burr, honeysuckle for bird's-eye maple and nectarine for mahogany swirl. Although the pictures show the sections reassembled, for easy reference, you should not reassemble them until you have finished graining all the sections and allowed them to dry.
3 Grain the walnut stripe figure (see pp. 114–5). This acts as a border and directs the eye towards the diamond feature.
4 Next paint the burr oak figure (see pp. 110–11). This provides a neutral dividing border between the strident walnut stripe and the featured central disc.
5 Use bird's-eye maple (see pp. 120-1) as a light-toned, anonymous backdrop.
6 Complete the composition by adding the central feature, a circle of mahogany swirl (see pp. 90–1). Seal and protect with two coats of mid-sheen polyurethane varnish.

MAPLE

Used in solid form from the Gothic period up to the seventeenth century, supplies of maple dwindled after the eighteenth century and it became prized as a decorative veneer. The colour of maple varies from pale silvery grays through creamy whites to the more pronounced golden-yellow tones of the matured timbers. In conjunction with the light, reflective qualities of the wood, extensive mottling produces a silky, lustrous quality in the surface.

Plain sawn timber shows a pleasing curly grain figure, and the darker, late growth rings provide an attractive contrast with the light overall tonal value of the wood. In radial sawn timber, silver cross-hatched rays are visible as fine flecks running parallel with the grain.

Although fiddle back, tiger figure, blister, landscape and burr cuts are commonly displayed on many elegant pieces of furniture, it is the 'bird's eye' variety that represents the most definitive maple figure. The bird's eyes, which appear as small shady hollows, are produced by a fungal infestation that in many cases invades the entire body of the tree. The surface may be likened to a road map, with the eyes representing isolated townships and the growth ring figures some complex inter-connecting road system. The figure passes, encloses but never dissects the eyes.

The natural material is prized for the clean, reflective quality of its surface. It is used extensively as a cabinet wood and combines well with the darker rosewoods and mahoganies on all types of furniture. Although the wood tends to be rather sophisticated, it has a warm appeal, and may be extensively represented by the grainer within the home without the surface becoming overbearing or strident. The unusual combination of lines, dots and mottles provides endless inspirational material for the production of fantasy variations.

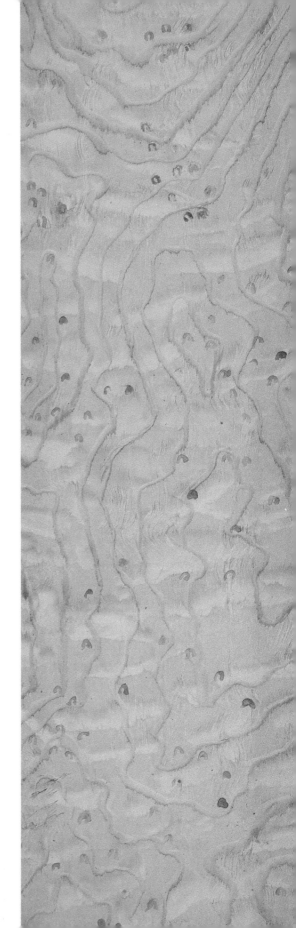

Bird's eye figure

1 Prepare a ground coat of eggshell finish paint in honeysuckle, slightly lightened by the addition of a little white artist's oil colour. Using a standard decorators' brush, apply this to a previously prepared surface (see page 62). Allow to dry for 24 hours. In order to prevent the water-based glaze from cissing when it is applied to the oil-based ground, scatter a small quantity of fuller's earth powder onto the panel and lightly rub it into the surface with a damp cloth. (Alternatively you can blend the powder into the glaze before it is applied to the panel.)

2 Using gouache colours, prepare a dirty-yellow water-based glaze by blending raw Sienna with small quantities of burnt Sienna and raw umber. Thin the glaze to transparency with water and apply it loosely to the panel, using a broad brush to create slight tonal variations.

3 Lightly soften the panel by striking the surface in all directions with a broad badger brush. This should produce a range of soft tonal transitions.

4 Using a crisp piece of plastic film, wipe out a series of mottles in the glaze. Displace the excess glaze towards the bottom edges of the mottles. The mottles should have a horizontal bias and should together produce a flowing, wavy pattern over the surface.

5 Using a large badger brush in a predominantly downward direction, soften the effect of the mottles. Blend the concentrations of glaze in the bottom sections of the mottles and merge them into the ground colour to produce the necessary tonal variations.

6 As the glaze begins to stiffen, drag an overgrainer through the mottles from the top to the bottom of the panel in a series of hesitating swerves. Work systematically over the panel, but vary the technique to avoid a repetitious effect. The deeper tones of glaze will show a variety of ragged, curly grain formations. By this time the surface will be almost dry.

7 Prepare a slightly deeper and more

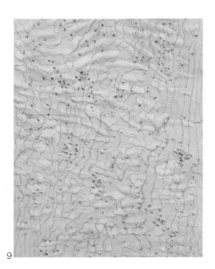

opaque version of the glaze in Step 2 from the original ingredients. Using an artist's pencil brush, paint in the growth ring figure. Soften as you proceed, using a small badger brush. The pattern should be similar to the lustrous rippling qualities that often appear on the surface of a pond. The newly applied glaze eats into the underlying colour and as you draw the small badger brush across the line a lighter band will be revealed on the inner side of the figure. In the natural wood these lines are continuous and you should work from the inside of the enclosed shapes, softening outwards.

8 Now dot the eyes onto the panel using the artist's pencil brush. Rotate the brush slightly as you paint in the eyes, to create hollow broken circles. Scatter the eyes casually over the surface of the panel. Sometimes they can be isolated and sometimes you can group them into small clusters. When you are satisfied with the appearance of the panel, allow it to dry out.

9 Complete the simulation by applying two coats of mid-sheen polyurethane varnish. Although the finished wood appears quite uniform, the figuring is quite complex and initially it is best to copy a sample to get the feel of the general patterning.

Note: This simulation is described in water colours, but it is also possible to carry it out using artist's oils. Although the simulation is completed in one continuous operation, there is considerable skill in judging the correct time to both distress the wet glaze and apply superimposed colour. As soon as the original glaze prepared in step 2 has been applied to the panel, it begins to stiffen. Trial panels should be prepared in order to investigate the different finishes that can be secured by varying the timings between the various stages. Bear in mind that water-based paints, such as gouaches and acrylics, dry much faster than oil-based types.

VARIATIONS

A The bird's eye figure makes a good background for a fantasy marquetry. Here, a light green version is combined with a darker green landscape or swirl figure.

B Yellow-orange eggshell paint has been used as the ground for this fantasy. First a transparent burnt umber glaze was brushed onto the surface, and the mottles were wiped out. Then, while the surface was still mobile, a deeper glaze composed of burnt umber and burnt Sienna was taken up on a fine artists' pencil, to create figuring.

A

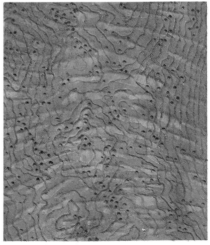

B

PINE

The collective name applied to the many varieties of white, softwoods that proliferate in vast forests in the Northern hemisphere is pine. The tree grows fast and straight and throws off many short, regular branches. When reduced to timber, the soft, porous grain structure is seen to be liberally scattered with many knots, discolorations and irregularities. The unpredictable structural performance and seasoning qualities of the timber traditionally restricted its use to all but the most utilitarian applications. However, with the dwindling supplies and increasing cost of the more substantial hardwoods, it has established itself over the last few hundred years as one of the most practical and popular woods for use in the home.

New wood is silky white to cream in colour. Protected with wax, it will mellow with age from light fawn through pale gold to deep tan. Paradoxically the very faults that define pine as a second-grade timber are those which give the wood its most pleasing figuring characteristics. Were it not for the regular branching formations, the tall erect stance of the tree would produce little in the way of figuring detail. Although there is a distinct tonal transition between the early and late growth rings, it is the knot formations that are primarily responsible for the many lively and elegant flame figures that are to be found liberally scattered throughout a typical sample.

Pitch pine Found in North America, pitch pine is primarily valued as a source of pitch and turpentine. It is robust, and is used as structural timber and as a cabinet wood. It is coarse textured and mid-brown in colour, with a distinctive graphic figuring pattern. The wavy lustrous qualities of the sap wood resemble Hungarian ash, while the heartwood, although more rounded, is not dissimilar to oak.

Pitch pine figure

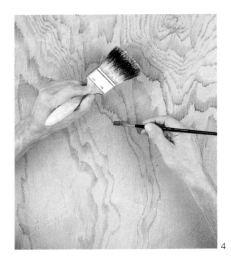

1 Mix a ground coat of eggshell-finish paint in cameo, slightly toned by the addition of a small quantity of black artist's oil paint. Using a standard decorators' brush, apply this to a previously prepared surface (see page 62) and allow to dry for 24 hours. Prepare a water-based glaze by blending burnt umber, burnt Sienna and raw Sienna gouache colours. Thin to transparency and apply quickly to the panel with a broad, flat brush. As you proceed, pounce a series of thin horizontal mottles into the surface. Allow the panel to dry out.

2 Prepare an oil-based glaze by mixing raw Sienna, burnt umber, burnt Sienna and black artist's oils. Add some transparent oil glaze, reduce with white spirit and, using a 2.5 cm (1 in) decorators' brush, brush-grain the glaze evenly over the entire panel.

3 Prepare another oil-based glaze by once more blending raw umber, burnt Sienna and burnt umber artist's oils. Reduce this with a little white spirit to produce a rather stiff glaze. The aim is to produce a glaze that sits on top of the underlying colour, as opposed to one that eats into it. Take up the glaze on a broad flat fitch brush and, using the italic quality of the brush, paint in the major lines in the grain. Soften with a small badger brush as you go along. Strike the figure outwards, from the centre of the heart, when softening.

4 Draw in the finer detail using a small fitch brush. Vary the softening technique to ensure variety in the pattern. Use the underlying mottles to suggest changes of direction.

5 Use a jamb-duster brush to introduce the characteristic ragged serrations in the figuring of pine wood. Drag the brush vertically through the glaze, once more pulling out from the centre of the heart figures. Although this wood is grainy, further texturing of the surface by, say, flogging would reduce its bright clean appearance. Leave to dry for 24 hours.

6 Seal with two coats of matt or mid-sheen polyurethane varnish.

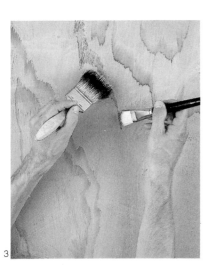

Yellow pine figure

1 Mix a ground colour by lightly tinting honeysuckle eggshell-finish paint with a small quantity of white artist's oil colour. Using a standard decorators' brush, apply this to a previously prepared surface (see page 62) and allow to dry for 24 hours. This example involves simulating two adjacent pine boards. Draw a vertical ruled pencil line down the middle of the panel and mask out the right-hand half of the board. Prepare a pale brown water-based glaze by mixing raw umber and raw Sienna gouache colours. Thin to transparency with water and using a flat brush, brush-grain the left-hand half of the panel to produce a series of fine, slightly undulating parallel lines. You should aim for an uneven distribution of colour which will ensure that the tonal variation of the panel echoes that of the natural wood.

2 Soften the left-hand section of the panel with a large badger brush. Do this by striking the surface across the axis of the grain with the brush. Allow the panel to dry.

3 Remove the masking and remask the painted section. Bring only a narrow strip of low-tack masking tape (say 3 mm/⅛ in) into contact with the previously painted surface to prevent the applied colour from being disturbed. Using the flat brush, brush-grain the right-hand side of the panel with water-based glaze, as described in Step 1, but work in a slightly different direction to provide a pattern that distinguishes this side from the left-hand side of the panel.

4 Soften the colour on the right-hand side of the panel, as described in Step 2. Again, remember that you are trying to produce the illusion of separate pieces of wood. Allow this side of the panel to dry and remove the mask.

5 The position of the knots has an important influence on the overall appearance of a piece of pine wood. Introduce them onto the panel now and they will serve as a guide for future figuring details. Prepare a mid-brown oil-based glaze by blending raw umber and burnt

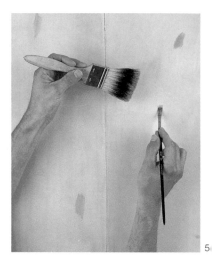

2

3

Sienna artist's oils with white spirit and a little scumble glaze. Using a small flat fitch brush, draw in the knot figures and soften them with a small badger brush, working mainly along the panel's vertical axis.

6 Paint in the figuring detail with the flat fitch brush and the glaze prepared in Step 5, using the position of the knots as a guide to the pattern. Remember to strengthen the composition by ensuring a clean separation between the contrasting figures painted on each side of the panel. Soften as you proceed with a small badger brush, being sure to work outwards from the heart's centre.

7 Using a deeper, more opaque version of the glaze mixed in Step 5, strengthen the knot formations and paint in their details with the flat fitch brush. Now allow the panel to dry out for 24 hours.

8 Darken the glaze you prepared in Step 5 by adding a small quantity of black artist's oil. Add a little transparent oil glaze and thin to transparency with white spirit. Using a 2.5 cm (1 in) decorators' brush, brush-grain the entire panel and distribute the colour unevenly to ensure a variety of tones.

Wipe out selected parts of the glaze by using the most suitable edge of a pencil eraser. Concentrate on the inner edges of the growth ring figures, the circles around the knots and the mottle formations that are generated by the knots.

10 Soften the wiped-out sections and toned areas with a small badger brush. Make sure that you retain and strengthen the individual identity of the two boards. When you are satisfied, allow the panel to dry for 24 hours.

11 A detail shows a typical knot and the softened edges of the grain pattern.

12 Seal and protect your work with two coats of matt or mid-sheen polyurethane varnish. The finished board shows that the broad grain patterns and knots of pine wood can be highly decorative. The knots can be selectively enlarged and the details varied.

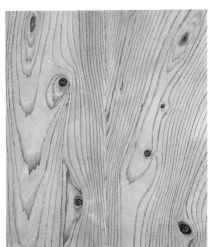

INDEX

ACKNOWLEDGEMENTS

Picture Credits

Adam Arbeid, Specialist Paint Finishes, Brighton: 22
Arcaid/Richard Bryant: 34–5; /Grainer: Sheila Sartin: 47; 63
Bolton Museum & Art Gallery: 28
Michael Boys Syndication: 20 (Insert), 34BL, 44–5
Camera Press: 48BL, 61, 69
Martin J. Dodge, Wincanton: 55
Macdonald/Orbis: /Photographer Jerry Tubby, Grainer Michael Snyder, Interior
Designer Dominique Lubar of I.P.L. Interiors: Cover, 23, 41TR;
/Photographer Martin Cameron, Grainer Peter Farlow: 9, 26–7, 36–7, 38–9,
41BR, 46, 50, 57–8;
/Photographer Jon Bouchier, Grainer Hannerle Dehn: 30 (insert);
/Photographer Susanna Price, Grainer Peter Farlow: 32–3;
/Photographer Jerry Tubby, Grainer Michael Snyder, Interior Designer
Francois Gilles of I.P.L. Interiors: 42–3;
/Photographer Dave King: 64;
/Photographer Susanna Price: 67, 81;
/Photographer Susanna Price, Grainer Henryk Terpilowski: 82–125;
National Trust: 10 (insert)
Pipe Dreams/Sheila Fitzjones P.R.: 40–1
J.M. Ratcliffe & Co, Southport: 6, 6 (insert), 20, 30, 72; /N.T. for Scotland:
18–19, 54
The Royal Pavilion, Brighton/Scala Publications Ltd: 8
Fritz von der Schulenberg: 25
Jessica Strang: 72 (insert)
Syndication International: 29
Weidenfeld & Nicolson Archives: 42BL
World of Interiors/Nadia Mackenzie/Tomkins: 49

Every effort has been made to trace the grainers and paint finish specialists,
and we apologise in advance for any unintentional omissions. We would be
pleased to insert the appropriate acknowledgement in any subsequent edition
of this publication.

Macdonald Orbis would like to thank Plotons
of Archway Road, London N19 for their
kind loan of equipment for photography, Michael
Snyder, IPL Interiors of 44 Fulham Rd, London SW3
and Farlow Boulter of 22 Grand
Union Centre, London W10 for their help in finding locations.